Prentice Hall Advanced Reference Series

Engineering

SELECTED TOPICS IN SIGNAL PROCESSING

Simon Haykin, Editor

Prentice Hall, Englewood Cliffs, New Jersey 07632

Library of Congress Cataloging-in-Publication Data

Selected topics in signal processing/editor, Simon Haykin.
    p.   cm.
    Includes index.
    ISBN 0-13-800947-3
    1. Signal processing—Congresses.   I. Haykin, Simon S.
TK5102.5.S4256 1989
621.38′043—dc19   89-15686
                                        CIP

©1989 by Prentice-Hall, Inc.
A Division of Simon & Schuster
Englewood Cliffs, New Jersey 07632

Printed in the United States of America
10 9 8 7 6 5 4 3 2 1

ISBN 0-13-800947-3

Prentice-Hall International (UK) Limited, *London*
Prentice-Hall of Australia Pty. Limited, *Sydney*
Prentice-Hall Canada Inc., *Toronto*
Prentice-Hall Hispanoamericana, S.A., *Mexico*
Prentice-Hall of India Private Limited, *New Delhi*
Prentice-Hall of Japan, Inc., *Tokyo*
Simon & Schuster Asia Pte. Ltd., *Singapore*
Editora Prentice-Hall do Brasil, Ltda., *Rio de Janeiro*

CONTENTS

PREFACE

A two–day Symposium on "Selected Topics in Signal Processing" was held at McMaster University, Hamilton, on May 25–26, 1987 under the auspices of the Communications Research Laboratory. The Symposium was organized as a contribution to McMaster's Centennial Celebrations. We were fortunate to have a slate of 12 distinguished scientists and engineers present lectures on various aspects of signal processing. The Symposium was attended by persons from both Canada and the United States. We were also honoured to have Vice–President L.J. King open the Symposium, and President A.A. Lee present a talk on "Research at McMaster".

This book presents a revised version of the Proceedings of the Symposium. Paper 1 is a reproduction of President Lee's talk on Research at McMaster. The remaining 10 papers of the book, Papers 2 through 11, are based on lectures presented at the Symposium. These 10 papers have been arranged so as to provide some sense of continuity. However, circumstances did not permit the inclusion of lectures given by Dr. D. Casasent, Carnegie–Mellon University, Pittsburgh, and Dr. S. Treitel, Amoco Production Co. Research Centre, Tulsa.

I would like to take this opportunity to express my gratitude to all the speakers at the Symposium. The efforts of those speakers who were able to contribute to the writing of this book are deeply appreciated. I am grateful to Paul Bauman and Anne Myers of the Communications Research Laboratory for their tremendous efforts to produce a camera–ready version of the book. Last, but by no means least, I am grateful to Bernard Goodwin of Prentice–Hall for publishing the Proceedings of the Symposium in book form.

Simon Haykin

1
RESEARCH AT McMASTER

DR. A. A. LEE*
President, McMaster University
Hamilton, Ontario

It is a pleasure to welcome you to McMaster during our 100th anniversary. We are enjoying the visits and attention of people like yourselves and finding it a good opportunity to review what has been achieved over the last years.

1.1 CRL

One of these achievements is the Communications Research Laboratory. It is a laboratory of which its members can be proud, for which the University is proud and grateful and which has made significant contributions to Ontario and the nation. It makes these contributions through the training of students, the publications of its research and the intimate involvement of the laboratory with industry.

In Ontario, the telecommunications industry is of major importance. It currently employs 93,000 people. It is faced with competitive threats, yet there are opportunities for significant growth in wealth and job creation. Over the next decade, telecommunications–based industries, according to one estimate, will grow throughout the world at an average annual rate of nine per cent. We think that a significant amount of that growth will be here in Ontario, thanks to our own Research Laboratory. We hope, for the sake of this country, that the pattern of the CRL's past successes will continue into the future and will add to the industrial base of this Province.

We are joined in that hope by twelve major companies and six governmental agencies with which the CRL has ongoing interactions. The research budget of the laboratory last year was two and one quarter million dollars. Over the past five years twenty nine Ph.D. degrees and sixty five Master's degrees have been awarded at McMaster in this area. All of this indicates a healthy enterprise. But there is no ground for complacency. Greatly increased resources are needed.

We are happy that fruitful ties between our own group and similar groups at Queen's, Ottawa, and Carleton are now being developed which will strengthen Ontario's and our own positions in this field. The sponsor for this new consortium–like coming together of researchers in the four universities along with industrial partners, is the Premier's Council in Ontario. The Council was created last summer to bring together in centres of excellence

---

*This is the text of a welcoming address by Dr. A. A. Lee, President of McMaster University, delivered at the McMaster Centennial Symposium on Signal Processing, May 25, 1987.

postsecondary institutions and the private sector to stimulate fundamental research; train world–class researchers; and encourage the transfer and diffusion of technology. Because the CRL is already strong and has private sector partners, it in a sense anticipates the thrust of the new centres of excellence being encouraged into existence by the Premier's Council. We are keeping our fingers crossed that this particular consortium will be one of the first six centres established.* If it is, there will be a large increase in the level of funding available now.

It is perhaps surprising, to those of us who are older, how quickly a major area such as this can develop. In 1970 the CRL did not exist at McMaster except in the imagination of Dr. Haykin. Since then it has become one of the major activities in our Engineering Faculty. Seventeen years is not a long time for a field to become established at a University and to make significant theoretical and practical contributions to the life of a country. Indeed if we think of institutions as analogous to human individuals, it is worth comparing the CRL to the length of time it takes to produce a productive scholar. We now expect most students to be about 28 years old before they begin to make any significant kind of contribution. So the CRL is, by individual human standards, both young and precocious.

1.2 RESEARCH AT MCMASTER

Although I have been at McMaster for twenty seven years, I still find it exciting to see new areas of expertise take root and flourish through breakthroughs and new perspectives in the intellectual areas that make up the map of a contemporary university. I have seen it happen before on quite a few occasions and yet it still is exhilarating to see the topnotch minds of faculty members and able students putting together innovative programmes of research and teaching in order to press the boundaries of what is known.

This exhilaration is no stranger to Engineering. Indeed, as most of you probably know, it was only one hundred years ago, in the year that McMaster was established, that engineers themselves became an organized professional body in Canada. That happened in 1887, when the Canadian Society of Civil Engineers was formed and elected as its first president, Thomas Keefer, the hydraulic engineer who designed the Hamilton water works. The Society later became the Engineering Institute of Canada. In 1887 in Canada, only McGill and the Ecole Polytechnic offered regular four year degree programmes in Engineering. Obviously, the study of engineering in Canada and the establishment of it as a research oriented activity in universities has seen remarkable growth since 1887.

For McMaster University as an institution, this process of establishing new areas of research began in a quiet way in 1942. It was then that McMaster began turning away from being an institution that primarily

---

*In June 1987, the establishment of the Telecommunications Research Institute of Ontario, as one of seven centres of excellence in the province, was announced.

transmitted knowledge to one which today spends much of its budget and many of its intellectual talents on the discovery of knowledge. To illustrate, in 1942 a combined arts and nursing programme was inaugurated which took five years to complete and was the formal beginning of health sciences at McMaster. Now, forty–five years later the wide range of health sciences on this campus makes our Faculty of Health Sciences a real pace–setter internationally as well as nationally, in research, in innovative educational programmes, and in providing leadership in health care delivery.

In the '50s, a third of a century ago, the change in the character of McMaster was signalled by two other major developments that have given this University important parts of its distinctive makeup: the nuclear reactor was planned and built under the leadership of Harry Thode, and John Hodgins was hired as the first director of engineering studies. Since that time nuclear science at McMaster has come to cover a broad range, from Dr. Brockhouse's internationally famous work in solid state physics to practical explorations of the mineral wealth of Canada through nuclear activation analysis. Around McMaster's early work in physics and chemistry the natural and biological science disciplines have developed in impressive ways. The Engineering Faculty, which is less than twice the age of the CRL, has for years held more research funding per faculty member than any other engineering Faculty in Canada. This is not a triumph of grantsmanship but a clear indication of leading–edge scientific and technological work by an impressive range of individuals whose achievements make this Faculty, closely linked with the Faculty of Science, potentially one of Canada's prime resources in reshaping its industries for success in rigorous international competition.

But this country is in deep trouble so far as Research and Development are concerned. The hard fact is that Canada's R & D culture is in a very rudimentary form. There are three issues that have to be addressed if we are to ensure continuing economic prosperity for this country as the world economy moves away from resource–based industries to knowledge–intensive ones: 1) the need to develop a long–term, applied research capability that is primarily industrially based; 2) the necessity of having effective basic research linked to this; 3) the improvement of our entrepreneurial capability to develop and market new products that should emerge from this dynamic.

At present the investment by Canadian industry in R & D, measured as a percentage of sales, is one of the lowest among developed western countries. Because there is no strong, long–term, applied research capability in Canadian Industry, it is almost impossible for the results of basic research in the universities to be used by Canadian–based industry in order to produce new products and services. Add to this problem the fact that the number of scientists and engineers per 10,000 of our work force is among the lowest among developed countries, and you begin to get a sense of how vulnerable we are.

Our basic research base is also being weakened. Until very recently universities like this one received almost no support for the indirect costs of research, with the result that we have had to drain the educational budgets to pay them. The federal research councils fund only the direct costs, with the result that both research and education are impoverished. Last year and this year, for the first time, in Ontario we have a partial remedy in some new earmarked funding for research leadership, 2.8 million for 1987–88. But set against an external research grants budget in excess of fifty million, this is only the beginning of a solution to the problem.

Back to McMaster's changes in recent decades. Concurrently with the strong developments in science and technology, from the 1950's onwards McMaster's Humanities and Social Science disciplines moved fully into the modern era of professional scholarship. Where earlier teachers of history or literature or economics were content to be intelligent readers and play their part as teachers transmitting knowledge of the past, their contemporary counterparts at McMaster, as elsewhere, are practising scholars, pursuing rigorous, difficult programmes of research inquiry and publication, for scrutiny by other professionals around the world. Less high–profile most of the time than their science, engineering, and health science colleagues, and not usually eligible for conspicuous sums of external research grants, the faculty members in Humanities and Social Sciences address questions of comparable complexity and importance and take on investigations crucial to our understanding of what the human race is and what it has done throughout history, as well as what it can now do in a period of human experience in which the world's populations and societies are rapidly coming together on a scale and in ways previously unknown.

The most recent addition to McMaster is a Faculty that expresses, in applied ways, both arts and science disciplines. This is the Faculty of Business or the "School of Business", as many prefer to call it. In its short history it has grown to the point of now badly needing a building of its own. Like most other parts of McMaster, it also is research–intensive. In fact, as measured by research activities, grants,and publications, it has become the most research–intensive School of Business in Canada. Here, too, there is a rich resource to help in the reshaping of Canada's economy, through the education of those who will manage the commercial enterprises, large and small, that are essential to the generation of wealth.

Co–extensive with the growth of fields, disciplines and numbers of students at McMaster, a growth which has been going on throughout the world, has been the increasing understanding of the purpose of all this activity: the discovery of the reasons for things. The search for reasons by those whose predecessors as faculty members spent their time transmitting what was known is now the defining characteristic of the university. The staggering growth in numbers and percentages of people attending university which has taken place over the last one hundred years has been accompanied by an enormous growth in fields of exploration and in the amounts spent on higher education and research. For example, in 1880 the entire university enrolment in all of Britain was 10,560 students, less than McMaster's full–time enrolment now. The participation rate of twenty to twenty–four year olds in Britain was six tenths of one per cent.

One hundred years ago, Cardinal Henry Newman (1801–90) was able to write disdainfully about the role of research in a university. For Newman and most of his contemporaries the function of a university was to help students come to know and appreciate what was already known, not to discover anything new. It is the special character of the contemporary university, however, that it regularly issues forth a stream of knowledge not known before to anyone in the world. This explosion of vast quantities of things coming to be known for the first time in the twentieth century is difficult to fathom or understand. To give you an idea of the sheer magnitude of it, we may recall that articles published in academic journals normally contain unique explanations of things, explanations that are persuasive to people accustomed to studying those same things. Many journals are published four times a year. At McMaster, despite pruning of acquisitions budgets over the last several years, we still subscribe to 14,000 journals. At the rate of four issues per year, that means that we receive roughly 150 journal issues every day, each of which is filled with articles explaining things hitherto unknown or misunderstood.

This kind of activity has not always been characteristic of the university. When I was an undergraduate at the University of Toronto in the fifties and when many of you were undergraduates, only a small proportion of the faculty was intensely engaged in research and scholarship. The majority were intelligent readers organizing and passing on the traditional lore of our culture. It is only in recent years that universities have begun to subject every element of the human and nonhuman worlds to intense examination in order to discover the reasons for things and to adapt those reasons to human purposes.

In university the relation of teacher and student is secondary to the authority of the subject being studied. At McMaster, unlike many universities, there is no such thing as "education" as a subject. There are only literature, physics, metallurgy, anthropology, finance, history, electrical engineering, neuroscience and similar subjects to be studied. The university does not ask if a man or woman is educated. It asks "What do you know?" "What do you do your research or scholarly work in?" This "what" is not mere content, to be used for personal or careerist development. What the man or woman knows is economics or philosophy or materials science or pathology or geology or whatever, and these are the organized forms of human knowledge.

The steadying, probing, complex, demanding discipline of the subject studied becomes an end in itself in proportion as the student matures. He advances from taking a course or two in a subject to being taken up in it. Many of our degree programmes are based on the principle that any genuine discipline can be used as a centre of knowledge, the mental distance the student moves out from that centre being his or her responsibility.

It is only in the highest levels of university education in graduate school at the doctoral level that methods are learned by which the search for the reasons for things can be carried out in a systematic and productive manner. In short, the student in the first four or five years of university learns to ask fundamental questions. It is only after years of study relating to particular questions that a subject becomes sufficiently understood that questions can even be properly focussed, so that the answers arrived at do produce new knowledge and understanding. This happens at about the time a Ph.D. is acquired. The relatively new requirement of the last thirty years that teachers in universities have Ph.D.'s reflects, then, a very new and different orientation of universities toward productive research.

While the desire to know runs through all the levels of education, primary, secondary, university, it is only at the point where new knowledge is produced and handed on to others, that the desire to know finally fulfills itself, at least insofar as this desire is held by modern people in technological societies. It is this activity of producing and passing on new knowledge that constitutes the modern world of research and university education. The university is a place that has made possible almost every detail of the man—created world in which we now live. It is this ongoing fundamental inquiry that has transformed us from a primal state of illiterate ignorance to one in which it is possible for large numbers of free people to seek answers to questions about the reasons for things and, having arrived at those answers, to use them to reshape our world more according to our desires.

The beginning of the life of reason probably occurs around age 18, but it takes carefully instilled habits of questioning, cultivated for many years, before it is possible for that restless questioning and thinking to result in the productive research of a faculty member. While this characteristic of knowledge is as ancient as man, the modern method of consciously and systematically exploring the reasons for things is one that has come to be only over the last three or four hundred years. As a common, institutional practice, it has come to exist in Canada only in our lifetime. Yet it is this sytematic inquiry into the reasons for things, and the passing on of those reasons to others, that characterizes the modern, technological world and the universities in that world.

It is for this reason that a conference of this sort is so appropriate to a university such as McMaster which is engaged in celebrating its centennial in 1987. McMaster is one of the most research intensive universities in Canada. This week marks the beginning of a meeting of over 5,000 other academics from across Canada and other countries who will be discussing their research, their ideas, and the directions in which their specialties are going. For the next two weeks we shall be immersed in a rich diversity of intellectual life. This large gathering has many intellectual connections with the ongoing daily life of McMaster. We celebrate that fact and we are delighted that you and many others are here with us to pursue complex questions and share your findings in an academically rigorous way.

2
VLSI TECHNOLOGY

PALLAB K. CHATTERJEE
Texas Instruments Incorporated
Dallas, Texas

The revolution that started in the late fifties with the invention of the integrated circuit has truly been the primary driving force for the explosive growth of most aspects of electronics technology. This has influenced the socio–economic structure of the world sufficiently to be compared with the effects of industrialization. Indeed, today's age has been heralded as the information age.

The backbone of the information age are the technologies for computational plenty. The technologies of the information age are:

- VLSI
- Symbolic processing, multiprocessing and system architecture
- Interface technology
- Advanced materials and device structure
- Mass storage.

As we evaluate the basic tenants of why integrated circuit technology has been so remarkably successful in orchestrating the explosive growth of the information age, one must point to the concept of device scaling. The general concept of feature size scaling as applied to the development of VLSI has brought with it an assured improvement in performance, density and cost. This is a remarkable combination since almost all other technologies have an increased cost associated with increased performance. An unprecedented improvement in manufacturing productivity in the semiconductor industry has made this possible.

In this paper, we shall briefly trace the major historical events of the last two decades that have brought us to the VLSI era. We shall then project the possible future direction of the application of VLSI technology to the continued development of the information age.

2.1 HISTORICAL PERSPECTIVE – THE EARLY DAYS

The integrated circuit revolution began in the late 50's with the invention of planar processing that enabled a fabrication of an active device, isolated from another and interconnected on the same substrate. In a quarter of a century, it has been possible to provide integration levels on a single chip of eight million components as demonstrated by the 4 Megabit dRAM shown in Fig. 1.

Tracing the history of this phenomenal growth leads us to two attributes of the semiconductor industry that are unparalleled in any other

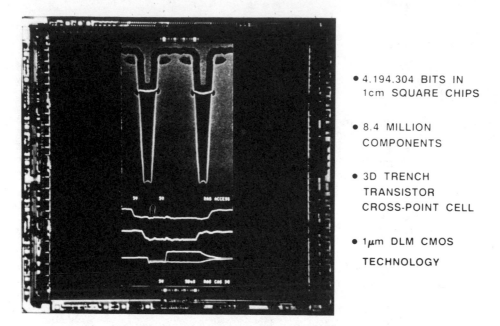

- 4.194.304 BITS IN
  1cm SQUARE CHIPS

- 8.4 MILLION
  COMPONENTS

- 3D TRENCH
  TRANSISTOR
  CROSS-POINT CELL

- 1μm DLM CMOS
  TECHNOLOGY

Figure 1   Photomicrograph of a 4 Megabit dRAM.

industry as of today.   The first is a concept of device scaling that allows
continuous miniaturization with attendant improvement in performance, density
and cost.   The second is the improvement in productivity through volume
learning that has allowed semiconductor component cost to decrease in the face
of rising costs and materials, equipment and labor costs.

The feature size scaling in integrated circuit technology has been a
primary force in its explosive growth.   The formalization of this procedure was
a result of work performed at IBM Research in the late 60's and early 70's.
The coordinated linear scaling of voltages, lateral dimensions, junction and
oxide thickness, and an increase in doping density, proposed on the basis of
this work was adequate in driving device technologies through the feature size
ranges from 7 μm in the dearly 70's to about 2.5 μm at the close of the
decade of the 70's.   The dominant device technology in the first decade (60's)
of integrated circuits was bipolar, and the circuit types of choice were TTL.
The advent of the MOS scaling theory, coupled with the development of clean
gate oxides, put MOS devices in prominence in the 70's.   The early
application was in metal gate pMOS which propelled the consumer business.
Silicon gate technology was key to the advent of higher speed nMOS.   NMOS
was the dominant technology for LSI applications in the 70's.   It opened up
the memory market and the microprocessor market.

2.2   THE ONE MICRON DISCONTINUITY

The prescriptions of the IBM linear scaling theory, however, ran into trouble
as feature sizes approached 1 μm.   There were three problems:

- Performance improvements in the down–scaled devices diminished
- Interconnect delays became larger than device delays
- Power management on–chip became difficult as we approached VLSI integration levels.

These were the major problems facing the world of VLSI in the early 80's. Let us examine the nature of each of these problems and the solutions that were put in place to continue scaling into the submicron regime. Most of these solutions involved structural and material changes, which we collectively label as the "one micron discontinuity".

The performance problems associated with the scaling of active devices resulted from two primary trends. First, circuit and system level noise margin requirements made it difficult to scale operating voltages. This resulted in increasing electric fields in the device. The large electric fields generated energetic "hot" carriers which could cause reliability and stability problems by getting launched into and trapped at the gate oxide. The second problem resulted from the increased resistance of the source/drain regions as the junction depths were scaled down. The resistance increase resulted in a negative feedback which limited the extrinsic gain of the device. These device structure problems are shown in Fig. 2.

In order to overcome these problems, two new concepts were introduced. The electric field was reduced by providing a graded junction near the drain. This has been called the lightly doped drain (LDD) structure. This results in a significant reduction of the drain electric field. In order to deal with the resistance problem, a self align metal silicide cladding process was developed. This cladding process allows the current to be carried in a refractory silicide layer over the source/drain and gate region. Fig. 3 shows this device structure.

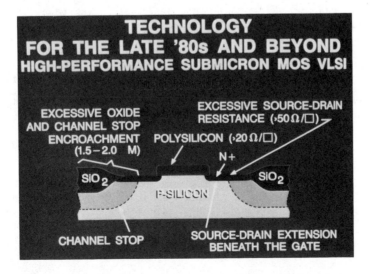

Figure 2   Device Structure Problems in Scaling Below 1 μm.

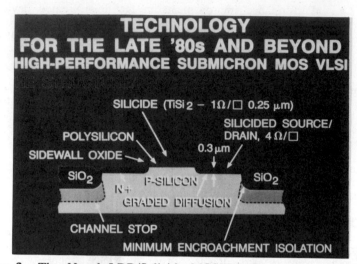

Figure 3   The Novel LDD/Salicide MOSFET for Submicron VLSI.

In addition to the device problems described above, the extent and quality of the isolation region between devices needed to be improved to continue to achieve improvements in packing density.   There have been many structural modifications required to achieve the isolation profile shown in Fig. 3.

Another structural change that occurred in the one micron regime addresses the problem of scaling memory devices – in particular, dRAMS. The storage element of a dRAM has a capacitor and an MOS transistor.   The charge stored on the capacitor represents the signal.   In scaling the storage cell, it is desirable to store a constant amount of charge while reducing the surface area of the cell.   This is possible in 3D structures where the walls of deep trenches can store charge.   Trench capacitor technology has been introduced at the 1 Megabit integration level.   At the 4 Megabit level, both the capacitor and the transistor can be integrated into the trench as shown in Fig. 4.   This transition into the third dimension allows the scaling of memory to continue.   The scanning electron micrograph of a trench capacitor dRAM cell is shown in Fig. 5, while Fig. 6 shows a totally 3D structure which puts the entire storage cell in the vertical dimension resulting in a near ideal cross–point memory cell.

The power management issues in MOS VLSI have been addressed by the use of CMOS circuit style where complimentary n–channel and p–channel devices are used in series to provide zero static power.

The combination of these structural changes (Fig. 7) required a major engineering and materials learning effort.   These changes need to be put in place at about the one micron feature size in order to enable the improvements in scaling to occur in the submicron regime.   This is referred to as the one micron discontinuity.

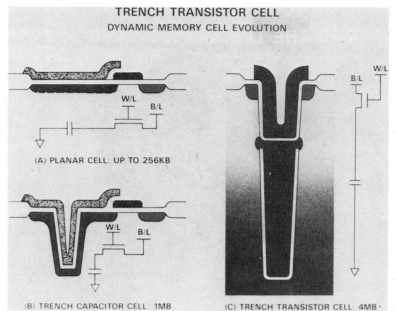

Figure 4   The Progressive Use of 3D Structures for VLSI Memories.

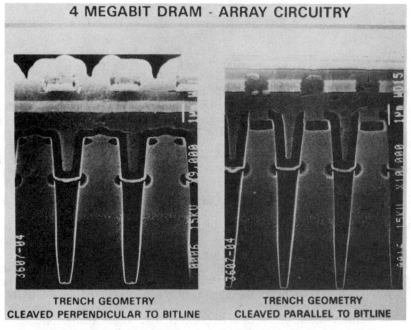

Figure 5   SEM Cross–section of a Trench Transistor Cell.

TRENCH TRANSISTOR CELL
MEMORY ARRAY, ISOMETRIC BISECTION

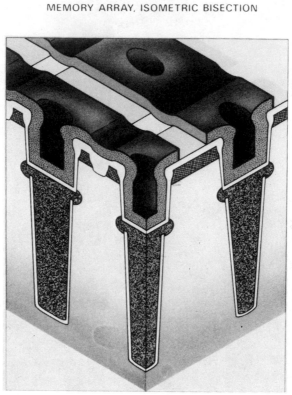

Figure 6   A Cross–point Trench Transistor Cell.

The outstanding problems in the ability to scale effectively hinge around the scaling of interconnects. Fig. 8 shows the capacitance per unit length as a function of feature size. As interconnect lines are brought together, mutual capacitance due to fringing increases. This results in a net increase in capacitance and coupling noise. This phenomena is a major problem and is the major focus of attention for technology development in the future.

2.3   CURRENT TRENDS IN SEMICONDUCTOR TECHNOLOGY

The advent of VLSI integration levels with its attendant cost/performance benefits has been able to stimulate the traditional uses of components, where a set of increasingly complex standard parts are made available to the electronics industry. We expect these benefits of the device scaling to continue into the next decade. The major challenges here are now in the manufacturing technology required to continue the improvement of cost and productivity.

An entirely new class of components has been made possible in the last few years, which are spawned by improvements in design technology.

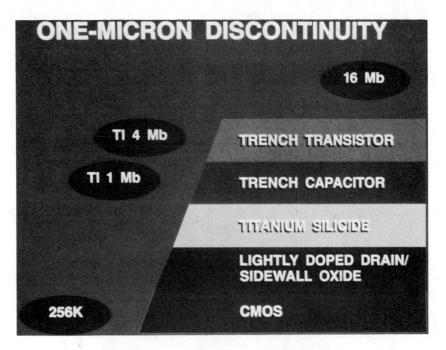

Figure 7   The One Micron Discontinuity.

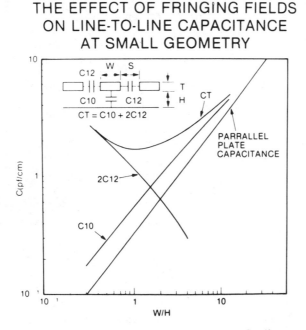

Figure 8   Problems in Capacitance Scaling.

This is a class of application specific integrated circuits (ASIC), which are chips that have been designed for custom applications. The key ingredient in this process is the active participation of the customer of the chips in the specification and even the design process. Design systems developed for this class of circuits are automated sufficiently to synthesize or compile the physical design based on a logical specification (Fig. 9). The chips are custom fabricated and the key index of importance to the customer is the turn–around time. As systems grow in complexity, it is expected that this class of devices will take on a major share of the logic market and allow the product definition task to primarily reside with the chip user. One may look at a chip design of this type to be an alternative to board level design, where the components are available in software and can be put together as a chip rather than a board.

The ASIC systems routinely used today provide the capability of integrating subcomponents in the MSI, SSI level. The future systems will aim toward providing LSI level subcomponents that can be used in ASIC style customization of ICs.

We believe that the enabling technologies to make true VLSI ASIC style designs possible are symbolic processing and object–oriented data bases.

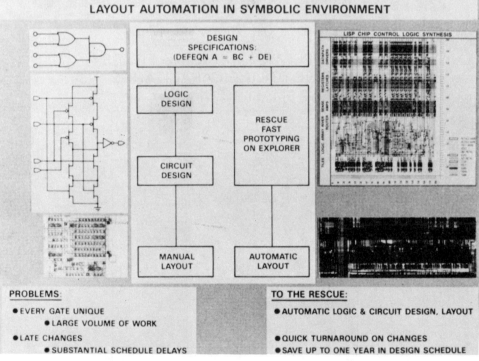

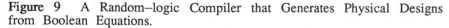

Figure 9   A Random–logic Compiler that Generates Physical Designs from Boolean Equations.

The ability to manipulate symbols, represent designs at various views and synthesize through constraint propagation are key to this class of design. The use of expert systems and heuristics is the only solution to many design problems. Most of the design problems are n–p complete in complexity.

Another major feature that is evolving in the design of ASIC chips is the ability to provide system level simulation and make design trade–offs at the highest level. Synthesis from the register transfer level of specification is being addressed. Perhaps future designs will be synthesized from behavioural levels of specification.

Fig. 10 shows a state–of–the–art 32–bit microprocessor developed at Texas Instruments, which typifies the directions and status of design systems today for VLSI chips. Most large chips require a significant amount of on–chip memory. This is best designed by hand to get good packing density. It is possible to parameterize the size of the memory for specific applications. The data path is a semi–automatic assembly of standard MSI level subcomponents. The control logic, which has the minimum structure and is the most difficult part of the design, was automatically compiled from Boolean equations. The theme in such a design choice is that structured or arrayed parts of the design can be optimized for packing density through a library of subcomponents, whereas the random logic and glue logic should be automatically synthesized.

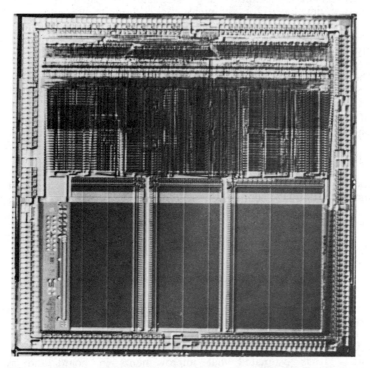

Figure 10   A 32–bit Microprocessor for AI Applications.

In order to be able to use the ASIC circuits in subsystems and systems, the design must meet functional, performance and reliability criteria. The design systems that are now available can guarantee functionality. Performance and reliability are typically not guaranteed in VLSI level designs. ASIC chips in the LSI and lower levels of integration have been able to reach a reasonable level of performance and reliability assurance by providing significant "factors of safety". In this strategy there are sufficient "guard bands" in time, voltage and power provided in the library to enable the chip to meet the customer's requirement. This style of design under–utilizes the intrinsic capability of the technology. As technology evolution becomes more expensive, this trade–off will not be as attractive. Further, many new reliability concerns are emerging in the submicron regime. Design systems that can allow performance and reliability constraints to be applied to the synthesis area (in addition to functionality constraints) will be required to optimally utilize submicron VLSI technology.

2.4 FUTURE DIRECTIONS

There is every indication that the improvements in performance and density using scaled VLSI will continue through the next decade. The technological problems associated with the scaling of the active devices down to below 0.5 μm are being researched today. The use of deep UV lithography for patterning of 0.5 μm features has been shown to be feasible. Plasma etching techniques for pattern definition are adequate for such fine features. The device structure developed for submicron MOS circuits are adequate for 0.5 μm geometry. The major research in the future will be in interconnect technology. Some improvements in isolation technology will also be required to achieve this density.

The next change in circuit technology will be the merging of bipolar and CMOS devices – BICMOS. The major technology imperative driving this change is performance and noise issues. BICMOS gates are quite insensitive to fan–out and are able to drive large loads with little degradation in speed. Thus BICMOS provides CMOS power and bipolar speed. Integration of these devices is a powerful way of improving performance at a given feature size. The high performance devices can open up many new applications.

The manufacturing technology required to provide quick turn–around, small volume, customized components will likely undergo a significant change compared to today's practice. The manufacture of many thousands of circuits requires factory management techniques that do not exist today.

The use of 3D devices for memory application will definitely increase. It is likely that logic structures can be integrated inside trenches to provide a whole new class of programmable logic arrays.

Perhaps one of the most explosive growth areas is 3D packaging. In the last two decades, there has been little progress in the technology of mounting chips on boards for use in systems. Many new packaging technologies that use the silicon chip directly on a circuit board are under development. It is estimated that improvements in board and system level packaging can lead to a density and performance improvement that is as dramatic as those that can be obtained from feature size scaling.

It is quite likely that device architectures that minimize interconnects will be invented in the next decade. This would be the most desirable solution to the total interconnect problem.

2.5 SUMMARY

We have discussed the evolution of IC technology in this paper. The phenomenal growth of the IC industry has been fuelled by improvements in productivity via the learning curve and the theory of device scaling.

The semiconductor industry is currently undergoing a transition from a standard component business to a customer specific component business. This transition is making possible a large number of new applications. The primary driver of this change is the improvement in design technology.

The improvements in semiconductor technology will continue at a similar pace in the next decade. The pacing item is almost totally based on the economics of the chip industry. A trend toward vertical integration is very evident. This will be the enabler to drive next generations of integration.

Technological advances in 3D device structures and 3D packaging are the key areas of growth in the next decade.

Overall, the semiconductor technology is well poised for continued growth in the future, at rates that are comparable to the past decade.

3
COMPUTATION, MEASUREMENT, COMMUNICATION AND ENERGY DISSIPATION

ROLF LANDAUER
IBM T.J. Watson Research Center
Yorktown Heights, New York

3.1 INTRODUCTION

There are three different, but related, fields in which we can ask about the minimal energy requirements, imposed by the laws of physics. The three fields are:

1. Computation
2. Measurement
3. Message transmission.

The distinction between computation and transmission is emphasized in Fig. 1. The basic logic step in a computer uses a strongly nonlinear interaction, between two or more digital input signals' streams, to produce one or more outputs. By contrast, in message transmission, we hopefully extract a copy of the message we inserted into the transmission channel. The only part of the computer which is like that is the memory. It is no surprise that communications oriented information theory is directly applicable to computer memories [1]. Computer memory, just like a communications channel, can be protected against a modest error rate by a little bit of redundancy.

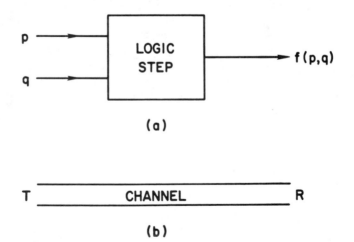

(a)

(b)

Figure 1 Logic step in (a) utilizes a nonlinear interaction between p and q. The transmission channel in (b) hopefully reproduces its input as the output.

Some mischief has been caused in the literature by an occasional over–simplified equivalence between the three fields listed above. The mischief is sometimes aggravated by dependence on a sloppy view of the measurement process, supposedly the oldest and best understood of these three fields. Questions about energy requirements for the measurement process resulted from concern with Maxwell's demon and, in that sense, are over a century old. Quantum measurement theory, furthermore, [2,3] is a highly developed subject. While much of this measurement literature is sensible, a totally satisfying correct and complete answer to Maxwell has only become available recently [4,5,6]. Indeed, much of the literature in this field does not really define *measurement*. Is any physical interaction between two systems, allowing one to influence the other, a measurement? If so, it certainly need not be a dissipative process. That conclusion contradicts the prevailing wisdom, which believes measurement requires energy dissipation. If not all processes coupling degrees of freedom are a measurement, what is the definition of measurement?

It turns out that a computer, as a closed system, is easier to analyze. The insights gained there, in the newest of our three fields, allow us to come back and settle the two other questions more definitively. We will argue that in all three areas an inevitable dissipation arises only from steps which discard information. In a final, and more speculative section, we will discuss the ultimate source of dissipation and noise; how it does appear in systems whose time evolution is governed by lossless and deterministic laws.

3.2   COMPUTATIONAL LIMITS

Information, whether in a computer, in a biological system, or on paper, is inevitably associated with physical degrees of freedom. This leads us to ask how computation is restricted by the laws of physics, and by the storehouse of parts available in the universe for assembly into a computer. While the inevitable physical embodiment of information seems to be widely accepted, today, one still occasionally finds statements of the following sort:

For instance, it is entirely outside our experience that a sequence of symbols on a piece of paper or on a tape, or held in someone's head, can ever act as the physical *cause* (as distinct from an instruction about what to do) of an entropy reduction in some entirely separate system.

taken from Ref. [7]. We intentionally cite this particular book, because in many other ways it is a perceptive and scholarly work, and we will use it as a source on several subsequent occasions. Having cited Ref. [7] we will, as a discursion, add some more comments stimulated by it. This book is largely devoted to a discussion of the subjectivity of entropy; is entropy determined by *our* lack of knowledge? Ref. [7] defends the objective nature of entropy, a quaint and outdated need. Surely in a day where we have an abundance of systems which carry out measurement under computer control, record the measurement, and then control a subsequent process, it must be clear that it is at most the existence of information which is relevant, and not whether it is located in a computer memory, or a human brain. Indeed, one can defend the proposition that information in a well understood and easily characterized system, which is accessible and measurable, is of far more significance than information in a brain, accessible only through reliance on an individual's communications. The same unnecessary emphasis on information held by human observers that Ref. [7] attacks, also occurs in discussions of the measurement process. Szilard's famous paper [8] is titled, "On the Decrease of Entropy in a Thermodynamic System by the Intervention of Intelligent Beings." It is, incidentally, an unnecessary reference to "Intelligent Beings;" Szilard's detailed text does not really return to this notion.

The quotation given above, from Ref. [7], causes us to ask: What about adiabatic demagnetization where we permit aligned spins to randomize, thereby taking up entropy from the environment? What is the difference between electron spins pointing up or down, and bits? The stubborn can, of course, take refuge in a distinction between information and its physical representation. We declare this to be an unverifiable and unfalsifiable distinction; it has no relation to science.

We add one more incidental comment, related to both Ref. [7], and to an enthusiastic review of the book [9]. Both items seem to be written without awareness of the fact that we can assign a non–vanishing entropy to a particular particle configuration, not only to the ensemble of which it is a member, and this entropy can be *large*, for a typical "random" (i.e., hard to specify) configuration. Bennett [5] has demonstrated that the algorithmic entropy, which measures the difficulty in specifying a configuration [10], is really the physical entropy.

In our concern with physical computer limits, we address the uncircumventable ultimate limits of computation, much as the second law of thermodynamics set limits on the performance of heat engines, and information theory set limits on the capacity of a noisy channel. We are not asking about the practical capability of real technology.

The study of ultimate computer limits as a serious field *with well-defined questions* has been in existence for more than a quarter century. While it has attracted limited attention, the number of contributors and papers is large compared to unity. We cannot summarize it here but only allude to some of its conclusions. A recent paper [11] can provide a start to the citation trail; we will also refer to a few of the papers in the field in our ensuing discussion.

Our summary of the field is best approached in a historical way. Almost from the beginning of the modern digital computer this was a subject for casual speculation. It is easy to connect a binary variable, in a computer, with a physical degree of freedom and thus with the typical thermal energy kT *associated* with such a degree of freedom. But *association* is an ambiguous notion. By 1961, it was understood [12] that it is the operations in a computer which discard information, that require a minimal and unavoidable energy dissipation. The required energy loss is of order kT per operation, the exact amount depending on the exact operation. Bennett [13] showed that computers do not really need to discard information. Such *reversible computers*, which use only one–to–one logical mapping at every step, require a great deal of extra equipment, or extra steps, and are not likely to see real use.

Reversible computers have been described in a variety of different embodiments, and by authors with varying viewpoints. The authors represent computer science, device physics, and quantum measurement theory. Classical (i.e., not quantum mechanical) reversible computers, in systems with friction and noise, have been described in particular detail. The friction in these systems is assumed to be proportional to the velocity of motion, as in hydrodynamics and electricity. The energy lost to friction can, therefore, be made as small as desired *per step*, by using a sufficiently low computational velocity. This is the same assumption that is implicit (and rarely stated) in the typical discussion of thermodynamic cycles. Such reversible computers can be given any desired immunity to noise *in the intentionally utilized degrees of freedom*. The preceding italicized phrase is used to distinguish against fluctuations which actually cause the hardware parts to suffer unintentional changes and act as a source of equipment deterioration.

Reversible computers with frictional forces proportional to velocity can have manufacturing tolerances, just as real computers do. The parts do not have to be perfect as long as they fall within a certain tolerance range. *Ballistic* classical computers have also been described [14,15]. These are systems without friction which carry out a computation with the kinetic energy given to them initially. These systems, however, have no tolerance for defects. The system must be perfect, without deviation from the specification. No friction or noise can be allowed. Furthermore, the moving parts of the system must be launched, initially, in a precisely specified manner, without any deviation. Thus, frictionless computers are pathological, and not realizable.

Finally, following the pioneering work of Benioff [16], quantum mechanical ballistic computers have been discussed. For further references and this author's evaluation of the literature, see Ref. [17]. Once again, these are pathological systems, without tolerance. Friction is essential to computation, a point we shall try to make plausible, here. If we want to use imperfect systems then we must discard the errors imposed by the defects; otherwise, the trajectory of the computation will deviate more and more from its intended path. But discarding the effect of errors is discarding information. And, as already pointed out, throwing information away requires energy dissipation.

Deutsch [18] has suggested that a quantum mechanical computer can be put into a linear superposition of initial states, each representing a separate digital input. The computations can then be carried out concurrently. If the right measurement is made at the end, it will provide information depending on all of these simultaneous computations, and effectively provide parallel computation on the superposed initial states. This requires a totally coherent quantum mechanical computation; no inelastic processes which destroy phase memory are allowed. This raises two questions. First of all, is such a long quantum mechanically coherent trajectory realizable? While we cannot anticipate all future inventions, it seems unlikely. The reader may be tempted to think of superconductivity; but superconductors at non–vanishing temperature and non–vanishing frequency are not dissipationless. Furthermore, all of the known proposals for superconducting computers involve intentional dissipation, switching from the superconducting state to the normal state in the presence of current flow. A more serious question, however, is whether dissipationless computation is desirable. As already indicated, dissipationless computers have no manufacturing tolerances, they must be perfect. Dissipationless computation is, in fact, not at all desirable!

When I was first exposed to Bennett's notion of reversible computation, it took me months to overcome my skepticism, despite the fact that my own work had, in places, come very close to an anticipation of Bennett's idea. Since then, however, the concept has become revisited by different investigators, with differing viewpoints. Despite that, there are occasional continued objections. These, typically, come from people who find that the concept violates their *perception* of general principles; the objectors do not tell us where the specific embodiments that have been invented go wrong. Most frequently, the objection depends on the joint suppositions that computation requires measurement, and that measurement requires dissipation. But all this is invoked without a careful definition of *measurement*.

The assumption is sometimes made that even if the internal operations in a machine can be reversible and can be carried out with minimal dissipation, this cannot be true for the input and output operations. The obvious answer to that: Yes, of course, these operations, like all others, can be done with great dissipation. But is the dissipation inevitable? Is there a reason why these input/output operations *have to be different* from those carried out inside the computer? Of course not. Some readers may want to respond: But at the output you *must* make a measurement, and that takes energy dissipation. We ask the reader with that inclination to suspend judgment until Sec. 3.4 has been read. But, in the meantime, ask yourself: What is a measurement? If it is simply information transfer, that is done all the time inside the computer, and can be done with arbitrarily little dissipation.

3.3. MODULATED POTENTIAL MACHINES

There are about six separate reversible computer inventions in existence. (The exact count depends on a subjective judgment about the minimum distinction required, in order that related inventions be counted separately). These were listed in Ref. [19]. Here we will describe only one of these schemes, adapting our discussion from that provided in Ref. [19]. In 1954 von Neumann filed a patent appliction [20] issued in 1957, and assigned to IBM. The same concept was discovered independently by Goto [21] in Japan. These proposals utilized parametric excitation of subharmonics, in nonlinear circuits, and that interesting history is discussed in some detail in Ref. [19]. The von Neumann—Goto scheme was adapted by Keyes and Landauer [22] to time—modulated potential wells. That work preceded awareness of reversible computation, but was the first paper showing that energy losses of order kT, per logic operation, were not just a lower limit, but could, in principle, be achieved. The paper also demonstrated that if computation is done sufficiently slowly, then any desired immunity to thermal equilibrium noise could be obtained. Likharev [23] later showed how the proposal of Ref. [22] could be modified to obtain a reversible computer. We shall now describe such a reversible computer, adding a few further refinements to Likharev's discussion. Likeharev's discussion invoked Josephson junction circuits instead of time—modulated potential wells, or parametric circuits, but these are all equivalent. We shall continue our discussion in terms of time modulated wells.

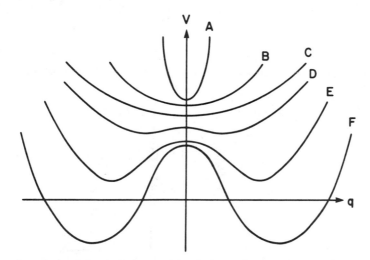

Figure 2   Potential changing with time.   Starts at A with a single minimum and ends up at F in a deeply bistable state, and then returns to A.   Relative vertical displacement of curves is unimportant, and selected for clarity.

Our basic device consists of a particle in a time dependent potential, shown in Fig. 2.   The potential starts as a monostable well with a single minimum.   The potential becomes deeply bistable and subsequently is brought back to the monostable state.   At the bifurcation point the well is very flat, and the particle is easily pushed one way or the other.   This biasing force is derived by coupling to other particles, which at an earlier time reached their deeply bistable state, as schematically shown in Fig. 3.   Fig. 3 illustrates a case where the three wells shown on the left have a majority of particles in the right hand half, and will push the particle in the well to be influenced to the right.   The right hand well, in Fig. 3, will, later on, when it is deeply bistable, influence other wells, in turn.   Springs are shown to illustrate the coupling mechanism in Fig 3.   That is somewhat symbolic; it is the relative displacements, away from the well center, that actually need to be coupled. For further details concerning the relative timing of the potential modulation, the reader is directed to Refs. [22]–[24].

The time–dependent potential of Fig. 3 can be produced, for example, by the periodic motion of charges to and from the (charged) information bearing particle under consideration.   Changes in interaction energy between the particle and the source of the potential variation do not require dissipation. Dissipation arises only when the information–bearing particle moves laterally, against friction.   Thus, to minimize dissipation we must invoke slow potential variation.   To be sure that the particles go as intended we must do two things:

(1) Modulate the potential slowly, so that we are never far from the time–independent Boltzmann distribution for the particle.

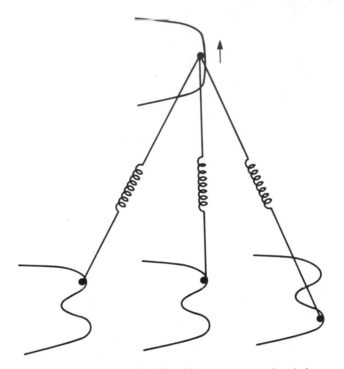

Figure 3   Three wells in deeply bistable state, on the left, coupled to the one about to undergo transition to bistability.

(2) Use large potentials, and strong springs, so that the probability of a particle left in the unfavored side, after passing through the bifurcation threshold, becomes very small.

The modulated potential machine has two blemishes when compared to some of the other reversible computer models, particularly Bennett's springless clockwork Turing machine [5].   One blemish comes from the fact that for a given design, i.e., given potentials and springs, there will be a small residual error probability determined by the Boltzmann factor, exp $(-\Delta U/kT)$, for residence in the unfavored well.   This error can, of course, be made smaller by going to larger forces.   This relative blemish may be more apparent than real; all reversible computer schemes have to invoke arbitrarily large forces somewhere.   Bennett's clockwork machine does it by invoking "hard parts." The second blemish: This machine has good noise immunity only when operated very slowly; otherwise, the Boltzmann factor, which provides that, has not had time to become effective.   The model also has an advantage when compared to other reversible computers.   It *is clocked* and moves through the computation at a predictable rate.

The original Keyes–Landauer [22] proposal was not a reversible computer.   After all, a lost vote as shown in Fig. 3, is lost information. Likharev [23] provided the necessary improvement, whose simplest embodiment is shown in Fig. 4.   Each well is simply connected to its private memory

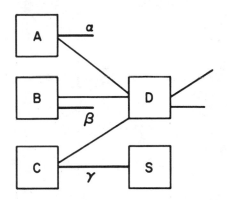

Figure 4 Wells, A, B, and C influence well D. Each well, however, is also coupled to a shift register, recording the history of the earlier states of that well. S is the first stage of such a shift register.

chain in shift register form, which remembers all the earlier states of that well. The coupling to this chain outweighs the other couplings. Thus, the information destruction associated with a lost vote, is eliminated. A more appealing alternative to Fig. 4 will be described shortly.

We do not intend to make this a design handbook for the modulated potential well; nevertheless, a few additional points need to be covered. The majority logic approach, schematized in Fig. 3, cannot produce a negation. (In the specific case of the original parametric scheme of Goto and von Neumann, negation is carried out very easily through a 180° phase shift, at the subharmonic frequency). To accomplish negation we use what is called dual–rail logic. We carry out all logic in duplicate versions. In one version, 0 is represented by a left hand well, in the other version by the right hand well. Negation is then accomplished by interchanging signals between the two branches. We will also need stages in which one (or more) of the inputs is fixed at 0 or 1, and the remainder variable. Fig. 5 shows how "or" and "and" are accomplished. Fig. 6 combines a number of such functions, employing dual rail logic, and shows how a family of needed logic functions can be constructed. Fig. 6 exhibits a reversible logic function, and avoids the need for the private shift registers of Fig. 4. In the reverse computation the coupling springs implicit in Fig. 6 will insure restoration of the left hand variables, p, q, $\bar{p}$, $\bar{q}$, to their proper values. We will not attempt to discuss that, here; it is best figured out by the reader.

3.4   INFORMATION TRANSFER, COPYING, AND "UNCOPYING"

Transferring information from one part of a computer to another, and making a second copy of existing information are normal computer functions. Thus, if the time modulated potential scheme can be used to build universal computers, it must be able to carry out these functions. It is apparent, for example, that we can build shift registers, in which information is simply pushed along. This is, already, a form of information transmission, and can be accompanied with arbitrarily little dissipation per transferred bit. From a single initial well

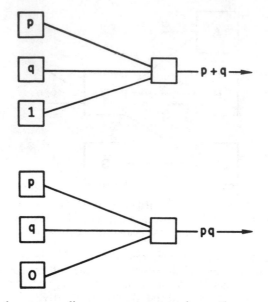

Figure 5   In the upper diagram a constant input 1 causes the majority logic to generate the logical "or".   The bottom diagram shows an "and".

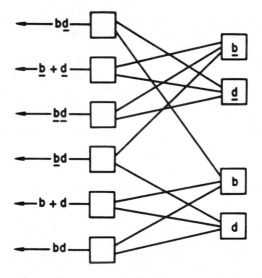

Figure 6   Execution of a set of simultaneous logic functions which eliminates the need for the shift registers of Fig. 4.

we can also control two subsequent wells, thus making two identical versions of the original bit.   To be quite sure that the point is made, we shall describe alternative schemes, using techniques which were not actually invoked in the previous section, but are closely related.

Fig. 7 illustrates a process for creating an additional copy of a bit. We start with a bit held in the left hand well shown in 7(a), ending up with the two positions of local stability right next to each other, separated only by a narrow impenetrable barrier.   Then in 7(b) we introduce a coupling spring, of the right length so that it can be fitted in, without stretching it or

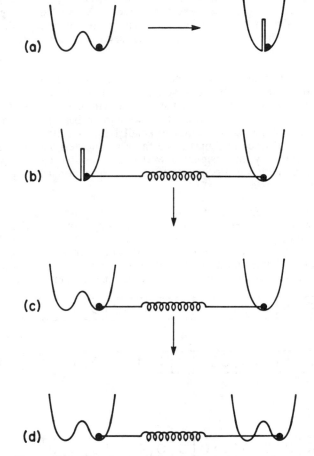

Figure 7   Reversible Copying.   The information to be copied resides in the choice of one out of two wells, as shown on the l.h.s. of (a). In (a) this bistable well is narrowed, bringing the two locally stable positions together. In (b) our narrowed initial well is at the left; the well which is to receive the information at the right, and a coupling spring connected.   In going from b to c the left–hand well is restored to its initial state.   In the transition from (c) to (d) the right–hand well is modulated as in Fig. 2, ending up in the deeply bistable state.

compressing it.   The left hand well in 7(b) has been compressed, as was
shown in 7(a), so that the same spring can be used for either bit position in
the left hand well.   (This is an alternative to other methods that have been
described in the literature, and the description given here may be a little too
simple for the careful reader.   For example, we have not allowed for the
possibly independent fluctuations in spring length. The effect of these can be
avoided by keeping the spring at T = 0 K, and only inserting it or removing
it between wells which have been made very narrow. Or, we could, instead,
invoke springs whose stiffness is modulated, rather than taking them in and
out).   We then go to 7(c), by modulating the initial potential well on the left,
and returning it to its original state, where the two valleys are clearly
separated.   Now the spring does exert a force.   After that we go to 7(d); in
this step letting the right hand valley, which is receiving the information,
become bistable, as invoked in the previous section.   We now have a second
copy of our initial bit.

To be a little more precise we add an aside. The well which is receiving the
information can, initially, be a well with a single minimum. Alternatively the
particle can be in a well—defined state of a symmetrical well, say the left hand
well. This difference is unimportant, we can go back and forth between these
two reversibly, i.e., with arbitrarily small loss. We do this by deforming the
well in the presence of a bias force favoring the designated side of the double
well. In this way the designated well in the bistable case is mapped into the
single minimum; we are not destroying information by mapping both wells
into the same state.

We can also carry out the inverse of the process described in Fig. 7.
We can start with two identical bits, and "uncopy" one of them.   We return
one of the copies to a standardized state, without actually discarding
information, by doing it via a carefully controlled interaction between the two
copies.   Instead of time modulated wells we can do very similar operations
using uniaxial magnets, and modulating their free energy as a function of

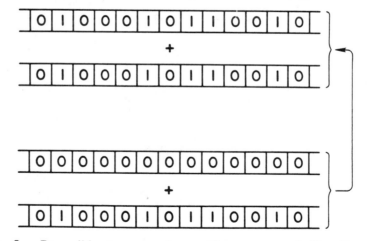

Figure 8   Reversible tape copying.   The process of Fig. 7 can be
used to copy the content of an information bearing tape onto a tape
with standardized bits.

magnetization by the use of a transverse field [5], or by temperature cycling through the Curie point.

Instead of copying a single bit, we can also, as shown in Fig. 8, copy a tape or string of bits. The copying must be done onto a tape which is in a known state, e.g. all "0" bits. If we start with a tape containing random information, and need to first erase or standardize the bits, that will require dissipation. That, however, in a system devoted to reversible operations, has to be done at most once. It is part of the manufacturing energy cost.

Fig. 9 shows an additional possible information transfer mechanism. We have a chain of bits shown at the top, with a "1", among a long string of "0" bits. We then copy the 1 onto its right hand neighbor. After that we "uncopy" the left hand 1. The result: The "1" has been moved,

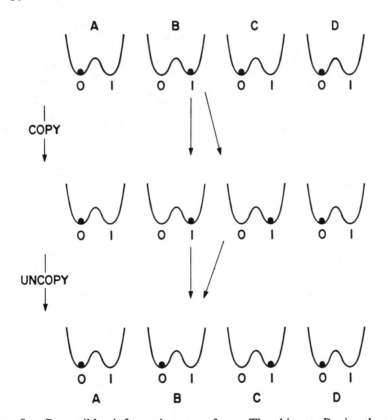

Figure 9   Reversible information transfer.   The bit at B, in the top row, carries information, and is surrounded by guard bits A and C which are intentionally at the "0" state.   The bit at B is then copied reversibly, leading to identical information at B and at C in the middle row.   After that the bit at B is uncopied reversibly, restoring the B bit to a "0" state, regardless of its earlier information bearing state.

reversibly, with arbitrarily little dissipation. For that particular scheme to work, we need guard bits with known content, to the right of an information bit. Thus, not every bit in this chain can carry independent information. But the information bits need not be separated by *long* intervals of guard bits.

3.5 ENERGY REQUIREMENTS IN MEASUREMENT

Maxwell, well over a century ago, posed his famous demon paradox [25]:

> "Now let us suppose that such a vessel is divided into two portions, A and B, by a division in which there is a small hole, and that a being, who can see the individual molecules, opens and closes this hole, so as to allow only the swifter molecules to pass from A to B, and only the slower ones to pass from B to A. He will thus, without expenditure of work, raise the temperature of B and lower that of A, in contradiction to the second law of thermodynamics."

This question has generated a huge literature. A key paper in this history is due to Szilard [26], available in an English translation [8]. Szilard's version of Maxwell's demon is illustrated in Fig. 10, adapted from Ref. [4]. A single molecule is presumed to be in a chamber; then (with arbitrary little dissipation) a barrier is inserted into its middle. A measurement is made, to see in which half the molecule is trapped. That information is then used to control the selection and motion of pistons, and the molecule does work against a piston. Finally, we return to the initial state. Szilard's paper makes no attempt to explore the "measurement" kinetics in detail. A more detailed analysis of Szilard's discussion and of other historical aspects of this field is given in Ref. [27]. Szilard, basically, asserts his faith in the second law of thermodynamics. Therefore, there must be a compensating entropy increase associated with the "measurement", so that no useful work is gained at the expense of thermal equilibrium energy, during the cycle. Szilard's discussion, however, isn't that definitive in placing the entropy increase. We know that in going from (g) to (h), in Fig. 10, we are discarding information, and this is associated with an inevitable dissipation, $kT\log_e 2$. That is just enough to save the second law, we need not look for a minimal dissipation elsewhere. That, however, was not clearly understood in Szilard's time, and most of the literature following Szilard went astray and assumed that the dissipation must occur in the transition between b and c in Fig. 10, i.e., in the step which transfers information from the molecule, to the machinery which subsequently controls the piston. Well known discussions, by Brillouin [28], followed by a refined version due to Gabor [29], invoke the fact that to "see" a molecule a photon must be used. In order to distinguish it from the surrounding black body radiation, the photon must have an energy $h\nu > kT$. This energy is assumed to be lost in the process. A great deal of later literature, of which we cite only a few examples, echoes these notions [30].

In retrospect, this acceptance of the Brillouin–Gabor view appears as one of the great puzzles in the sociology of science. If someone proposes a method for executing the "measurement," i.e., the transition from b to c in Fig. 10 which consumes a certain amount of energy, why should we believe that the suggested method represents an optimum?

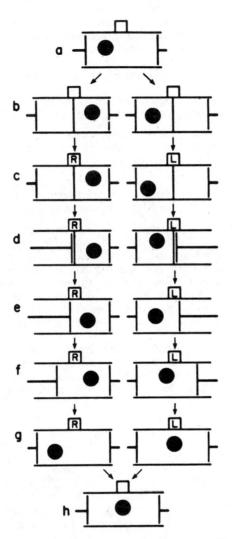

Figure 10   Szilard's version of Maxwell's demon.   The transition from a to b involves the dissipationless insertion of a barrier.   Between b and c a measurement is carried out, leaving information about the molecule's location in the (L,R) register.   In d a piston has been brought in, from a side controlled by the register.   Between d and e the initial barrier is removed.   In f the molecule does work against the receding piston, arriving at the initial volume in g.   Then the (L,R) register is reset in the transition to h.

Indeed Bennett has now shown [4] how the "measurement" can be done with arbitrarily little energy dissipation and without moving the molecule out of its initial half of the cell.   Bennett's invention of a specific detailed measurement method, applicable to Szilard's version of Maxwell's demon is welcome, and gives a finality to the discussion which was missing. Nevertheless, in general terms, it should not have come as a surprise.   After all, we have already seen that copying of a bit does not require minimal dissipation, and it is a similar information transfer process.

Fig. 11 characterizes the information transfer, or measurement, process. A system S and a measuring system M_o are coupled and evolve together for a limited time.   After that, they move apart, or are decoupled.   In the process, the measurement system has been changed from a standard initial state M_o to a modified state M' which relates to the state S.   As stated long ago [12]: *The mere fact that two physical systems are coupled does not in itself require dissipation.* The same point is emphasized by Peres [31].   Finally, of course, we need to reset M' to M_o, for the next use of the meter. Here we are discarding information; *this is the dissipative step.*   We could, of course, in principle reverse the whole system, S' and M', and return to S and M. That would be a resetting of the meter which could be done with arbitrarily little dissipation.   But this is not the typical measurement process; it would serve no purpose, it would undo the measurement, without utilizing it.   We could try to circumvent this by copying the output M', before reversing the measurement.   But that only postpones the dissipation at the expense of extra equipment. At some later time we will need to erase the copy.

If the measurement system is sufficiently damped compared to the time over which the object to be measured changes, then a continual resetting process is not needed.   The thermostat controlling a home heating system is an example; it does not have to be continually reset to a standard temperature.

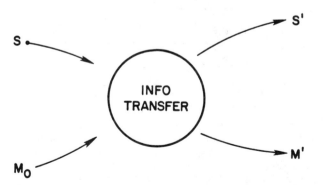

Figure 11   A system S is coupled to a meter.   The meter is in a standardized initial state M_o.   After a coupling is established and then removed, the meter is left in a state M', providing information about S.   The final state of the measured system may change to a new state S', but such a change is not essential.

The point we have just made is not new, but not widely understood. These concepts were discussed in detail in Ref. [5], and much more cursorily in Ref. [32]. A perceptive quantum mechanical version of this discussion has been provided by Zurek [6].

I have been careful to limit the discussion to the classical measurement process. The quantum mechanical discussions [3] present a complex diversity of viewpoints; we hesitate to add much to that, without more thought and conviction. Nevertheless, we add some reactions, clearly qualifying them as casual. The quantum discussions include some remarkable tales about Schrödinger's cat and Wigner's friend. Some of the greatest figures in the development of quantum mechanics present us with almost mystical beliefs. Ref. [2] presents us with many of these quotations, and quotes Bohr [33]:

> In such terminology, the observational problem in quantum physics is deprived of any special intricacy and we are, moreover, directly reminded that every atomic phenomenon is closed in the sense that its observation is based on registrations obtained by means of suitable amplification devices with irreversible functioning such as, for example, permanent marks on a photographic plate, caused by the penetration of electrons into the emulsion. In this connection, it is important to realize that the quantum–mechanical formalism permits well–defined applications referring only to such closed phenomena.

Wigner's views are summarized [34]:

> Wigner reasons that an observation is only then an observation when it becomes part of "the consciousness of the observer" and points to "the impressions which the observer receives as the basic entities between which quantum mechanics postulates correlations."

Is all this commotion about macroscopic objects and about human observers essential? The late twentieth century is a time when we can view, and perhaps soon place, single atoms with a scanning tunneling microscope. We can do spectroscopy on single atoms cooled by a laser beam. We have learned a great deal about molecular biology, and even have proposals (admittedly on a playful conceptual level) for applying such methods to universal computation [5,35]. As discussed in Sec. 3.2, we have a whole array of proposals for computation with systems which are Hamiltonian and quantum mechanical; admittedly, these systems have some problems. We routinely manufacture and measure devices which are on the borderline between macroscopic and molecular scales [36,37]. We have come far since the governor of a steam engine displaced the need for human observation. Admittedly, the more delicate modern experiments, such as the Scanning Tunneling Microscope, still involve an array of very macroscopic backup apparatus. But the burden of proof has shifted: Those who invoke the need for macroscopic apparatus in measurement, or go to the even more extreme extent of invoking human observers, must present a clearer case for their choice. Indeed, all the modern experiments presented at a recent conference on quantum measurement [3] point in one direction: all dynamical processes are controlled by the same quantum mechanical laws of time evolution. We can, of course, never be entirely certain that Ohm's law holds on the dark side of the moon, unless we have been there to test it. In the same way, we can allow continued speculation that quantum mechanics will fail somewhere [38]. I would not want to bet my own money on that. That, of course, does not imply that the currently perceived exact form of quantum mechanics is in its final form; indeed, one direction for further refinement is suggested in Sec. 3.8. Wheeler has suggested that human communication is an essential part of the formulation of the laws of physics [39]. I do not want to endorse this, neither do I want to challenge it; it is a worthwhile and stimulating suggestion. I have a much more limited objective here, which includes an objection to the need for human observers *in every single measurement*.

Thus, I believe that the discussion of the classical measurement process that has been presented, can be applied directly to the quantum mechanical case, in agreement with Zurek [6].

3.6 COMMUNICATIONS

In this Section we will question some of the most commonly accepted and quoted conclusions of Communication Theory or Channel Capacity Theory. The basic points of Shannon's contribution [40] are not our target. Shannon's general expressions which relate channel capacity to the probabilities for received messages, in terms of transmitted messages, are not in question. The problems arise when more specific physical models are invoked, and their generality overinterpreted. Shannon understood these limitations. Later workers have been less cautious. Quoting from Shannon: (with italics added for emphasis)

> An important *special case* occurs when the noise is *added* to the signal and is independent of it (in the probability sense).

In that case, Shannon finds the well known result

$$C = W \log_2 \left[1 + \frac{P}{N} \right] \; . \tag{1}$$

C is the channel capacity, W is the bandwidth, P the average received power, and N the average noise power. Thermal noise, for a classical transmission line with additive equilibrium noise, is given by $N = kTW$. Eq. (1) yields a maximum for C/P, at small P, given by

$$C = P/kT \, \log_e 2 \; . \tag{2}$$

This suggests that at least $kT\log_e 2$ energy per transmitted bit is required, though it is not clear that this has to be dissipated. As in the case of Szilard's contribution, discussed in Sec. 3.5, much of the problem in the literature has arisen from an excessively narrow interpretation of a pioneering thrust. Ref. [41] gives a typical development of this sort. We intentionally cite a contribution which is, in fact, more perceptive and stimulating than most.

Eq. (1) has an obvious physical reasonableness. N determines, in some rough sense, the distance between distinguishable signals. The larger N, the fewer the number of possible distinguishable messages. But communication does not need wave propagation. Mail is communication. Communication, furthermore, need not use degrees of freedom in which the effects of noise are added linearly to that of the signal, in the transmission process and in the detection process used at the receiving end. Furthermore, even if channel capacity expressions describe the energy required in the message, does it have to be dissipated?

The easiest way to see that Eq. (2) is of limited applicability is to consider information transmission by transport of a high density reel of tape, or disk [15]. It can be moved at velocities close to the velocity of light. The kinetic energy in the transport is not dissipation; it can be recovered at the receiving end. The reader may object: But you didn't do anything with the tape at the receiving end. That is correct and makes the key point: Whether dissipation is required, or not, depends on a detailed analysis of events at the receiving end. The fact that particles, rather than waves, can be used to transmit information has been recognized in Refs. [42] and [43], but does not seem to be widely appreciated in most of the channel capacity literature.

We have already seen, in Sec. 3.5, that information can be transmitted along chains of coupled particles in time dependent bistable wells, with arbitrarily little energy dissipation. These are, of course, nonlinear systems, with a local stability to the information bearing states. The effects of noise and of the intentionally applied forces do not superpose in a simple linear way. We could, alternatively, have used sine–Gordon chains [44] as the basis for a communications link, transmitting a succession of kinks. The details of that scheme, however, would require considerable auxiliary discussion, and do not seem essential, here.

We will supplement the discussions in Sec. 3.4 with one more example, illustrated in Fig. 12. There are two pipes, again with frictional forces proportional to velocity. A ball in the top pipe is a "0", in the lower pipe a "1". Below the pipes we show a moving periodic force field, or "buckets," which pace the motion of the information. (Moving periodic wells,

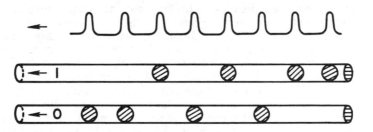

Figure 12 Binary information carried by balls in pipes. The selection of pipe determines the choice between 0 and 1. The particles are moved along the tubes by a traveling potential, with "buckets," shown at the bottom.

and the error probability caused by noise inducing a particle to fall into a well behind the intended one, were discussed in detail in Ref. [45]). Clearly this is an information transmission system; and the dissipation per transmitted bit can be made as small as desired by slow transport. At the ends the particles can come out of, or go into, information handling systems of several types. These could be the modulated potential wells, as discussed in Sec. 5, or could be Bennett–Fredkin–Turing machines, as discussed in Ref. [46]. Over the long run we may not want to accumulate physical balls on the receiving side. We can copy the received information, and then uncopy the information in the originally transmitted balls. Then we can return the physical balls in a standardized "0" state. Or, we can return the balls with their information content, and uncopy at the transmitting end. Needless to say, if we ever want to destroy a bit of information, then a minimal dissipation $kT\log_e 2$ is required.

But as in computation and measurement, *it is only information destruction* which requires minimal energy dissipation.

Note that the system of Fig. 12 not only avoids minimal energy *dissipation*, but also the energy *content* of the information stream can be very low. Let us assume for the moment, that the barriers between successive bit positions are less penetrable, quantum mechanically, than the barrier between pipes. We also assume that each pipe is narrow enough so that only the lowest transverse quantum–mechanical state in the pipe is occupied. The excitation energy to the next higher transverse pipe state, which has a wave function with a node *in the pipe*, is easily made large compared to kT. Now the ground state for a ball in a given "bucket" will be a wave function which is symmetrical *between the two pipes*, and has no node in the barrier between the pipes. If our particle is in one pipe or the other it is not in this ground state, but is in a superposition which includes that state and the next higher state which is anti–symmetrical between the two pipes. But this essential energy elevation can be made as small as desired by use of a sufficiently impenetrable barrier between the two pipes. Note, incidentally, that if leakage between the two pipes occurs largely through tunneling, then perturbation by the environment can actually reduce the tunneling rate [47], rather than act as a source of noise–induced activation over a classical barrier.

We have utilized particles in nonlinear force fields for our low–energy dissipation schemes. It will be interesting to learn if wave propagation schemes allow similar possibilities. In fact the modulated potential device of Sec. 3.3 started out over thirty years ago as a proposal utilizing bistable nonlinear resonant circuits [20,21]. Nonlinear wave propagation effects require high energies, and are not a likely *practical* route to communication with low dissipation. Indeed, the obvious analog of our pipes are optical fibers, and do not require nonlinearity [48]. But then what is the equivalent of our moving buckets of Fig. 12? Possibly, physical cavities which transport photons. Alternatively, traveling electro–optic modulation fields, that use nonlinear effects can be utilized. But once the trapped photon or packet of radiation arrives, it is less clear what we can do with it. Similarly at the launching end; our library of reversible computer inventions do not yet include any onethat utilize photons as the information vehicle. Furthermore, frictional losses for a particle are minimized by slow motion. On the other hand, if we transport an electromagnetic signal inside a slightly lossy moving cavity, the longer the transport takes, the larger will the loss be. These questions about nonlinear boson channels deserve further attention.

3.7   QUANTUM MECHANICAL COMMUNICATION

The theory of quantum channels, their capacity and energy requirements, is manifested in the literature through two, somewhat unconnected, branches. The invention of the laser, followed by two early papers [49] on quantum channel capacity, led to a highly developed body of mainstream literature in the field, of which we can cite only a few samples [50]. These papers are characterized by a great deal of valid concern with the nature of the apparatus at the receiving end. They are also characterized by a single–minded assumption that communication uses electromagnetic signals, or their close equivalent, and not the physical transport of mail or magnetic tapes. We will not comment further on this branch of the literature. The other "branch" is manifested through a set of papers [42,43,51–54] which pay little attention to the receiving apparatus, and little attention to the more widespread other branch [49,50]. This second branch tends to proceed by direct physical enumeration of the states of the transmission channel. It is more perceptive in the sense that *some* of these papers [42,43] do admit to the possibility that particles, and other sorts of distinguishable states, can be used to transmit information. The papers in this second branch are variable in their content and detailed approach, and by no means in agreement. Levitin [53] for example, refers to "energy used," whereas Pendry [43] is aware that the energy content of a message need not, inevitably, have to be dissipated. Indeed, Pendry [43] is probably the only channel capacity paper, aside from Ref. [15], which understands this point. Undoubtedly, the equations in Refs. [42,43,51–54] are manipulated correctly, and all of these papers have some range of applicability. Nevertheless, we question their implied generality, and assemble some assorted and incidental comments, in the following paragraphs.

Levitin, in Ref. [53], tells us that: *Brillouin (1960) showed through a number of examples that the minimal energy per natural unit of information is kT...* Note the ambiguous language: No attempt is made to distinguish between energy contained in the signal, and energy which has to be dissipated. Bremermann [52] tells us that with an energy E_{max} we can transmit, at most,

$h^{-1}E_{max}\ln(1 + 4\pi)$ bits/second. But this assumes a single transmission channel; we can clearly do better if we are free to allocate the energy among several independent transmission channels acting in parallel [55]. Additionally, Ref. [52] invokes our Eq. (1) as its starting point, and we have already pointed to the limitations of that. Bekenstein [42] starts by discussing a maximum information or entropy, for a given energy, in a given volume:

$$S/E \quad < \quad 2\pi R/hc$$

where R is the radius. This came out of Bekenstein's black hole work, and quite likely has its appropriate range of applicability. It has no applicability in the ordinary thermodynamic sense. The relationship dS = dQ/T tells us that at very low temperatures we can get arbitrarily high entropy changes, for very little energy input. Deutsch [56] makes this same point. Thus, if we have an ensemble of systems mostly in their ground state, and only a few in the first excited state, we get a great deal of entropy, for very little energy. Similarly in communications: A very rare photon (or fire alarm signal), with many intervening empty time slots, can carry a great deal of information through its timing. In Ref. [42] Bekenstein argues for his energy limit by counting the number of states available, up to a specified energy, in some typical systems, e.g. particles in a well or in a harmonic oscillator. If I wanted to compress a great many bits, however, with low excitation energy, into a given volume, I would not resort to such systems. It would be best to simulate the approach of digital memory, and use arrays of bistable potential wells, of the form invoked in Secs. 3.3 and 3.4. With a high barrier between the two wells the "0" (say l.h.s. occupation) and the "1" (say r.h.s. occupation) can be very close to the ground state. There is also no lower limit to the spatial extent of the states involved, if we can invoke sufficiently high barriers. We are, of course, here concerned only with the excitation energy required by information storage or transmission, and not with the total rest energy, including that of the apparatus. This latter energy seems to have been the object of concern in Ref. [57]. But in analogy with other similar fields, such as the efficiency of thermodynamic cycles, or channel capacity theory, we ignore the energy of the apparatus and other manufacturing costs.

While Pendry's discussion [43] seems to be the least flawed, it also leaves some points unanswered. Pendry finds the state of maximum entropy flow in a channel subject to certain constraints. His expressions involve an *information temperature*, T, but it is not really clear what physics determines this temperature. It is *not* the noise temperature of the channel material: Pendry's information carrying entities do not come into equilibrium with the channel walls. Pendry's temperature, instead, seems to characterize the choice of messages at the transmitter. Additionally Pendry's electrons are apparently uncharged; their Coulomb charge does not enter explicitly into their behavior.

3.8   ULTIMATE SOURCE OF NOISE AND IRREVERSIBILITY

The time evolution of closed, conservative, and predictable systems is the central realm of physics. Here, we include quantum mechanics under the label *predictable*; despite the occurrence of probability in quantum mechanics a quantum mechanical Hamiltonian time evolution causes no increase in entropy. Quantum mechanics, per se, does not cause systems to drift apart in phase space. For example: Two free particles, with a slight separation in momentum, can keep that particular separation indefinitely. By contrast, macroscopic phenomena are typically dissipative. How do we go from the conservative system to a satisfactory description of friction? The literature has a myriad of papers on the subject, and we can cite only a minute representation of these papers [58]. Many of these papers try to solve a psychological problem; they try to make the author, and sometimes even the reader, comfortable about the inconsistency. The most common approach is illustrated in Fig. 13. We consider a central closed system and then admit that this is not the universe. Effects from elsewhere perturb our system, and are the source of fluctuation and friction. For calculation purposes it is an effective approach. Our discussion, here, is adapted from Ref. [37].

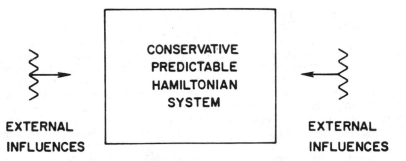

Figure 13   Transition from conservative Hamiltonian time evolution to dissipative and noisy behavior.   This is, most typically, explained by taking the central system under consideration, whose kinetics is described explicitly, and letting it in turn be subject to external influences.

Nevertheless, there is something basically unsatisfactory about the view symbolized by Fig. 13.   The boundaries of the system are selected arbitrarily. If we moved the boundaries further out, then the sources of friction and fluctuation would become part of a conservative, predictable, system.   In this section we want to point to a more basic approach.   Typically, we have viewed macroscopic systems as lossy, e.g. in viewing the rotation of a tire on the road.   We do not, however, talk about the friction seen by an electron in a molecule.   Is this dichotomy justified?   If it is, where is the boundary between macroscopic and microscopic?   In recent years there has been a great deal of investigation of very small conducting structures, intermediate in size between a molecular scale, and that which is obviously macroscopic [36,37]. This concern with systems of intermediate size has forced us to face questions more explicitly.   Here, we will not allude further to these solid state discussions, but invoke more broadly familiar situations.

We are really asking about the distinction between energy storage, i.e., a reactance, and energy dissipation, as in a resistance.   Resistances and reactances can be nonlinear and time–dependent.   When applying a time dependent voltage or field, the time dependence of the current is not a good indicator of the distinction between a resistance and reactance. Consider a length of electrical transmission line, terminated at its far end by a short circuit, or else left open.   This line will look like a resistance over a limited period of time, shorter than the round trip for the wave along the line.   But it is clearly an energy storage device, not a dissipative one.   An unterminated and semi–infinite transmission line does act like a resistor; the input energy just keeps moving away from the input.   But, even in that case, we can turn the energy flux around by applying a short across the line, at a point the signal has not yet reached. (The short, of course, must be applied at a predetermined time, or else controlled by a signal moving along a faster line. The latter is only possible if the original line has an effective geometrical velocity less than c, either because it is a wiggly line, or else loaded with dielectric material).

The use of an infinite transmission line, or the equivalent use of an unlimited array of linear oscillators, with the oscillators capable of absorbing energy from the system under consideration, are common devices [59] for simulating frictional effects in quantum mechanics. But, as we have pointed out, it isn't really quite enough to show that the array of oscillators can take up energy. To be a genuine model of dissipation, it must be shown that the energy *cannot* be retrieved. We admit that the use of coupled oscillators is an effective quantum mechanical calculational device, and only wish to point out that it is not, necessarily, a rigorous representation of dissipation The crucial distinction between resistor and reactor: Can we retrieve the energy? We now want to point out that such a distinction can be made without invoking the picture of Fig. 13, and in this connection become speculative.

Consider, for example, a simple pendulum, at its lowest point. Give it a kick. The energy can be retrieved by applying the same impulsive force when the pendulum returns to its original position. Note that this procedure does not require a pendulum which was initially at rest; we can be dealing with a thermal ensemble of pendulums. (However, if we need to allow for the dependence of the period on the total energy, due to the nonlinearities, it becomes much more complex). Our energy extraction procedure requires knowledge of the period. Furthermore, it requires knowledge that we are dealing with a simple pendulum. If our pendulum were coupled to others, then on its return to its original point of departure, it would no longer have all its initial kinetic energy. In fact, a generalized pendulum example involving coupling and nonlinearity immediately shows us that we must know the original system in detail, and must be able to calculate its subsequent motion, to be able to devise a method for extracting the energy. If the system is chaotic, following its motion and extracting energy is not ruled out, as long as we accept the prevailing notion that arbitrarily accurate measurement and arbitrarily precise calculation is possible. To follow chaotic motion, and predict the status of the system, with a specified accuracy, requires a calculational effort which grows exponentially with time, but that is not *impossible*, under our most common viewpoint, in physics.

Thus, if we assume that we can characterize our initial system completely, can calculate its subsequent time evolution to any required degree of accuracy, and can build the equipment needed to extract the energy, then energy put into a finite Hamiltonian system can always be extracted, and there is no dissipation. All of the assumptions we have made, however, are open to question, if the system has a good many degrees of freedom, and particularly if we have let a lot of time elapse since the energy was put into the system. Most clearly, however, as emphasized in Refs. [17] and [60], we believe that arbitrarily demanding computation is unlikely to be available, in *principle*, and is not just as a matter of the computing center's budget.

In conclusion: Even in a finite Hamiltonian system, unperturbed by external influences, energy retrieval cannot be implemented, unless the system is very simple and the elapsed time short. There is a need for a much more definitive theory, which describes quantitatively how much energy can be recovered with what probability and how the energy recovery diminishes with the complexity of the system. We are, clearly, still very far from that. While chaos, by itself, is not a source of unpredictability or irreversibility, it obviously does cause the need for computation to grow very rapidly with elapsed time. With 10^{23} molecules in chaotic collisions, it may not take many collisions to make an explicit tracking of the microscopic configuration impossible, in principle [61].   There has been some tendency in the literature to assume that chaos, by itself, provides an explanation for fluctuations.   That cannot be, chaos arises out of *deterministic* equations of motion.   Furthermore, chaos can be differentiated from real noise through the fractal dimension of the attractor, and other means [62].

We also remind the reader about the intimate connection between fluctuation and dissipation.   The preceding discussion has stressed dissipation, but the same influences which we have invoked as a source of difficulty in energy retrieval, also cause fluctuations to appear.   Fluctuations, after all, manifest themselves as a separation, with time, of supposedly initially identical ensemble members.   Inevitable difficulties in characterizing the exact system through measurement would, for example, cause initially indistinguishable ensemble members to behave differently later on.   Limitations, *in principle*, on the amount of computation would in turn limit the accuracy of physical laws [11,17,63].   This again, would act essentially as a noise source [46] and the motion of initially supposedly indistinguishable ensemble members cannot be guaranteed to evolve identically.   There is one aspect, however, that arose in the discussion of reversibility and dissipation that does not have an obvious and immediate counterpart in the treatment of fluctuations.   We alluded to the possible difficulty of assembling the equipment needed to extract the energy, and that does not seem to have an obvious parallel relevant to fluctuations. Indeed, the question: *What is reasonable equipment to invoke for energy extraction* has a depth, and we have only alluded to it superficially.   Consider the unterminated transmission line, where we were prepared to transport shorting bars, at relativistic velocities, to a section far from the initial end.   Is this really the same "system", as the initial unmodified simple line?   The "equipment realizability" question seems, however, to be the least relevant difficulty, because it only appears in the discussion of energy retrieval, and not in the discussion of fluctuations.

We have stressed the irretrievability of energy, *in principle*.   There is a much more readily achieved set of circumstances where energy is still retrievable in principle, but very difficult and unlikely, in reality.   Is it reasonable to call that dissipation?   Do we even need to know?   Is the distinction between stored energy, and energy which is not retrievable in principle, really needed?

We have, in a limited way, come full circle.   First we learned that discarding information requires energy dissipation.   Now we are suggesting that energy dissipation is the consequence (in part) of insufficient information to allow energy retrieval.   Thus, there is almost an equivalence between dissipation and an information loss.

Our explanation of irreversibility is not orthogonal to many older discussions.   Maxwell already stated [64]:

> Dissipated energy is energy which we cannot lay hold of and direct at pleasure, such as the energy of the confused agitation we call heat.   Now confusion, like its correlative term order, is not a property of things in themselves, but only in relation to the mind which perceives them.

The first sentence summarizes our view; it is unfortunately accompanied by a second sentence with its unnecessary reference to the "mind".   Indeed, even the first sentence is already misleading by use of "we", and should have, instead, stated, " ...which cannot be gotten hold of and directed by external intervention..." Ref. [7] is, as already stated, largely directed to the question: Is entropy subjective or objective? Is it personal knowledge that is relevant to the definition of entropy? As already stated, this seems to be an outdated debate. Information does not need the human mind for its existence. Information, whether it is held in a biological system, or elsewhere, is inevitably physical. We cannot, here, review Ref. [7] and the many eloquent scientists quoted in it, in detail. We touch only upon one more aspect of the book, a chapter devoted to *coarse graining*. Coarse graining is a set of techniques for taking the Liouville equation, of a closed and conservative system, which does not allow for entropy increases with time, and fudging it, to let that happen. Our point: The fudging is not needed. The ultimate version of the physical laws of motion is limited in their accuracy, and already contain something which is very similar to an external noise source.

Up to this point we have discussed the ability to retrieve externally supplied energy. As emphasized by Richard Liboff [65], irreversibility is also manifested in systems which have had no energy supplied through "terminals." Consider a gas in the left half of a vessel, restrained by a baffle. Take away the restraining baffle and let the gas expand into the larger volume. The gas, with its initial energy, according to a simple physical chemistry view now has a larger entropy; an irreversible event has occurred. Has it? Can the expansion be undone?

It is easiest to answer the question about reversing the chain of events quantum mechanically. If we take the complex conjugate of the many–particle wave function, then all particles will reverse their velocity, and the original time evolution will be undone. Unfortunately, taking the complex conjugate is *not* a physical operation. Physical operations must be represented by unitary transformation. Taking a complex conjugate is not a unitary transformation; *it is not even a linear operation*! This argument does not actually prove that there is no other way of chasing the particles back, but it seems unlikely.

Next let us consider the entropy. Here we need not distinguish between the classical case and the quantum–mechanical one. Let us assume, at first, that the system is truly isolated and Hamiltonian. Thus, we are assuming rigid vessel walls; we are not dealing with environmental influences as shown in Fig. 13. A Hamiltonian time evolution will provide no entropy increase; after expansion our original phase space has been mapped into a *subset* of the larger phase space which in turn corresponds to all the states with the given energy in the larger volume. From this viewpoint no irreversible event has taken place. In the quantum mechanical case this conclusion is at variance with our preceding argument involving the complex conjugate. This demonstrates the ambiguity of the word *irreversible*.

What we have just discussed in terms of phase space also follows from Bennett's discussion [5]: equating algorithmic entropy with the real physical entropy. The algorithmic entropy of the initial configuration consists of the number of bits required to define this state. Most random gas particle configurations will not allow an abbreviated description, as is possible for a crystal. We simply have to describe the configuration by specifying the state of each particle. A subsequent precisely specified Hamiltonian evolution will not change the algorithmic entropy. Instead of specifying the final configuration we can equally well specify the initial configuration. It is, after all, not really the length of the program which interprets the specification that counts; only the number of bits in that specification.

Consider the case where, during or after the expansion, we allow some effects due to the environment as shown in Fig. 13. This supplies the needed coarse–graining to let us populate the whole phase–space of the larger volume, and permits the entropy to increase as expected from elementary considerations. On the other hand, as suggested earlier, if the Hamiltonian evolution is sufficiently complex, then the needed coarse–graining can be supplied by the limited precision of the laws of Hamiltonian time evolution; external noise is not needed.

Can we undo the *classical* motion leading to the expansion, if we accept the ordinary interpretation permitting unlimited precision in the laws of time evolution? Do we have a way of turning around all the particles simultaneously (or in some way which allows for the lack of simultaneity)? Presumably we can anticipate all the positions and velocities, at some specified instant. Then we insert suitably oriented and suitably placed reflecting barriers, or apply the required reversal force to each particle in some other way. Is such a multitude of events realizable? The world is not classical, and this author has little confidence that motion can be followed with arbitrary accuracy. Therefore, it hardly seems worth asking the remaining question whether 10^{23} little fingers are available to manipulate the particles.

3.9 SUMMARY AND CONCLUSION

We have discussed minimal energy requirements for computation, measurement and communication. The newest of these areas, the computational process, is the simplest, because the computer is most nearly a closed and completely describable system. We discussed computation with emphasis on the modulated potential machine as an example of *reversible computation*, and with emphasis on the conclusion: There are no unavoidable lower limits on the energy dissipation per step. This includes transfer of a bit from one device to another, which is a form of communication. It also includes the creation of a second copy of a bit, which is a form of measurement.

In the discussion of energy required by measurement, Szilard's 1929 analysis of Maxwell's demon was a major step. Szilard reaffirmed belief in the second law, and that the measurement process, in some overall sense, requires energy dissipation. Szilard, however, did not pin down the exact source of the dissipation, within a measurement cycle. We now know, from analysis of the computational process, that resetting of the meter requires energy dissipation, and for Maxwell's demon this is enough to save the second law. Authors following Szilard, however, did not understand that, and looked for dissipation in the step in which information is transferred from the object to be measured to the meter. Brillouin, Gabor, and others found dissipative ways of transferring information, and without further justification, assumed that they had discovered a minimally dissipative process. It is one of the great puzzles in the sociology of science why this obviously inadequate argument met with wide and uncritical acceptance. Only in recent years have clearer discussions emerged, and these are not yet widely appreciated. We summarized the analyses by Bennett, augmented by Zurek.

In the communication link we face a similar history. Shannon understood that the linear case where noise is added to the signal, and is independent of it, is a special case. Later authors have been less perceptive and have tended to make interlocking assumptions; Information transfer is done by waves, is done in linear systems, and the energy in the wave must be dissipated. Physical transport of bistable systems violates these unnecessarily restrictive assumptions. We argued that in communication, as in our other areas, inevitable minimal dissipation arises only when information is discarded.

In our final section, and one which is admittedly speculative, we addressed the question: How can we explain the transition from conservative microscopic dynamics to dissipative processes? It is suggested that the limited precision available, *in principle*, in calculating the behavior of physical systems limits our ability to retrieve energy from supposedly conservative systems. This can be regarded as the ultimate source of dissipative processes.

3.10   REFERENCES

1.   J.A. Swanson, IBM, J. Res. Dev. **4**, 305 (1960).

2.   J.A. Wheeler and W.H. Zurek, eds. Quantum Theory and Measurement (Princeton Univ., Princeton, 1983).

3.   D.M. Greenberger, Ed., New Techniques and Ideas in Quantum Measurement Theory (Annals. N.Y. Acad. Sci., 1986).

4.   C.H. Bennett, "Demons, Szilard's Engines, and the Second Law," Sci. Am. **255**, 108 (November 1987).

5.   C.H. Bennett, Int. J. Theor. Phys. **21**, 905 (1982).

6.   W.H. Zurek in Frontiers of Nonequilibrium Statistical Physics, G.T. Moore and M. O. Scully, eds, (Plenum, N.Y. 1986) p. 151.

7.   K.G. Denbigh and J.S. Denbight, Entropy in relation to incomplete knowledge (Cambridge Univ. Press, Cambridge, 1985).

8.   A translated version can be found in Ref. [2] p. 539.

9.   A. Parsegian, J. Stat. Phys. **46**, 431 (1987).

10.   G. Chaintin, Sci. Am. **232**, 46 (May, 1975); G. Chaintin, IBM J. Res. Dev. **21**, 350 (1977).

11.   R. Landauer, Phys. Scripta **35**, 88 (1987).

12.   R. Landauer, IBM J. Res. Dev. **5**, 183 (1961).

13.   C.H. Bennett, IBM J. Res. Dev. **17**, 525 (1973).

14.   E.Fredkin and T. Toffili, Int. J. Theor. Phys. **21**, 219 (1982).

15.   R. Landauer, Int. J. Theor. Phys. **21, 283** (1982).

16.   P. Benioff, J. Stat. Phys. **29**, 515 (1982); Phys. Rev. Lett. **48**, 1518 (1982).

17.   R. Landauer, Found. Phys. **16**, 551 (1986).

18.   D. Deutsch, Proc. R. Soc. Lond. A. **400, 97** (1985).

19.   R. Landauer, in Der Informationsbegriff in Technik und Wissenschaft, O.G. Folberth, C. Hackl, eds. (R. Oldenbourg, Munchen, 1986) p. 139.

20.   J. von Neumann, Non–linear Capacitance or Inductance Switching, Amplifying and Memory Organs. (U.S. Patent 2,815,488).

21.   E. Goto, J. Elec. Commun.Engrs. Japan 38, 770 (1955).

22.   R.W. Keyes and R. Landauer, IBM J. Res. Dev. **14**, 152 (1970).

23.   K.K. Likharev, Int. J. Theor. Phys. **21**, 311 (1982); K.K. Likharev, S.V. Rylov, and V.K. Semenov, IEEE Trans. Magn. **21**, 947 (1985).

24.   R.L. Wigington, Proc IRE **47**, 516 (1959).

25.   J.C. Maxwell, Theory of Heat, 4th ed. (Longmans Green, London, 1871) p. 308.   See also M.J. Klein, Am. Scient. **58**, 84 (Jan.–Feb.,1970).

26.   L. Szilard, Z. Phys. **53**, 840 (1929).

27.   C.H. Bennett, IBM J. Res. Dev. 32, 16 (1988).

28.   L. Brillouin, Science and Information Theory (Academic Press, N.Y., 1956) Chapt. 13, p. 162.

29.   D. Gabor, in Progress in Optics, Vol. I, E. Wolf, ed. (North–Holland, Amsterdam, 1961) p.109.

30.   F.T.S. Yu, Optics and Information Theory (Wiley, N.Y., 1976) Chapt. 5, p. 93; A.F. Rex, Am. J. Phys. **55**, 359 (1987); J.D. Barrow and F.J. Tipler, The Anthropic Cosmological Principle (Clarendon, Oxford, 1986) see pp. 179, 662; also see Ref. [7] p. 2.

31.   A. Peres, Am. J. Phys. 42, 886 (1974); Phys. Rev. D **22**, 879 (1980).

32.   R. Landauer, in Sixth International Conference on Noise in Physical Systems, P.H.E. Mejier, R.D. Mountain, F.J. Soulen, Jr., eds. (Nat. Bur. Stand. (U.S.), Publ. 614, 1981), p. 12.

33.   Ref. [2] p. 3.

34.   Ref. [2] p. 769.

35.   C.H. Bennett and R. Landauer, Sci. Am. **253**, 48 (July, 1985).

36.   Y. Imry in Directions in Condensed Matter Physics, Memorial Volume in Honor of Shang–Keng Ma, G. Grinstein, G. Mazenko, eds. (World Scientific, Singapore, 1986) p. 101; S. Washburn and R.A. Webb, Adv. In Phys. **35**, 375 (1986).

37.   R. Landauer, "Electrical Transport in Open and Closed Systems," Z. Phys. B, 68, 217 (1987).

38.   A.J. Leggett in Ref. [3] p. 21.

39.   J.A. Wheeler, in Problems in Theoretical Physics, A. Giovanni, F. Mancini, and M. Marinaro, eds. (University of Salerno Press, 1984) p. 121; J.A. Wheeler, in Frontiers of Nonequilibrium Statistical Physics, G.T. Moore and M.O. Scully, eds. (Plenum, N.Y. in press); J.A. Wheeler in Ref. [3] p. 304.

40.   C.E. Shannon, Bell Syst. Tech. J. **27**, 379 (1948); ibid. p. 623.

41.   H. Marko, Kybernetik **2**, 274 (1965).

42.   H.B. Bekenstein, Phys. Rev. D. **30**, 1669 (1984).

43.   J.B. Pendry, J. Phys. A: Math. Gen. **16**, 2161 (1983).

44.   M. Büttiker, and R. Landauer, in Nonlinear Phenomena at Phase Transitions and Instabilities, T. Riste, ed (Plenum, N.Y. 1982) p. 111.

45.   R. Landauer and M. Büttiker, Physica Scripta **T9,** 155 (1985).

46.   R. Landauer, Ber. Bunsenges **80**, 1048 (1976).

47.   W.H. Zurek, in Foundations of Quantum Mechanics in Light of New Technology, S. Kamefuchi, ed. (Phys, Soc. Japan, 1984) p. 181.

48.   J. Rothenberg, private communication.

49.   G.J. Lasher, in Advances in Quantum Electronics, J.K. Singer, ed. (Columbia Univ. Press, N.Y. 1961), p. 520; ibid. J.P. Gordon, p. 509.

50.   C.W. Helstrom, Quantum Detection and Estimation Theory, (Academic Press, N.Y., 1976); F.T.S. Yu, Optics and Information Theory (Wiley, N.Y., 1976); C.W. Helstrom, J.W.S. Liu and J.P. Gordon, Proc IEEE **58**, 1578 (1970); R.O. Harger, ed., *Optical Communication Theory* (Dowden, Hutchinson and Ross, Stroudsburg, 1977); J.R. Pierce, E.C. Posner, and E.R. Rodemich, IEEE Trans. Inf. Theory 27, 61 (1981); Y. Yamamoto and H.A. Haus, Rev. Mod. Phys. 58, 1001 (1986); B.E.A. Saleh and M.C. Teich, Phys. Rev. Lett. **58**, 2656 (1987).

51.   J.D. Bekenstein, "Communication and Energy," preprint.

52.   H.J. Bremermann, Int. J. Theor. Phys. **21**, 203 (1982).

53. L.B. Levitin, Int. J. Theor. Phys. **21**, 299 (1982).

54. D.S. Lebedev and L.B. Levitin, Information and Control **9**, 1 (1966); W.G. Chambers, J. Phys. A: Math. Gen. **14**, 2625 (1981).

55. R. Landauer, and J.W.F. Woo in Synergetics, H. Haken, ed. (Teubner, Stuttgart, 1973) p. 97.

56. D. Deutsch, Phys. Rev. Lett. **50**, 631 (1983).

57. J.D. Bekenstein, Phys. Lett. **46**, 623 (1981).

58. B. Gal–Or, Science **176**, 11 (April, 1972); S. Watanabe, Phys. Rev. **84**, 1008 (1951); J. Mehra, ed. The Physicist's Conception of Nature (D. Reidel, Dordrecht–Holland, 1973); A.H. Kritz and G. Sandri, Phys. Today **19**, 57 (Sept. 1966); L. van Hove, Physica 21, 517 (1955); M.M. Yanase, Annals of the Japan Assoc. for Philosophy of Science 1, 131 (1957); B.L. Holian, W.G. Hoover, and H.A. Posch, Phys. Rev. Lett. **59**, 10 (1987). See Ref. [7] for many citations.

59. A.O. Caldeira and A.J. Leggett, Ann. Phys. (N.Y.) **149**, 374 (1983); A. Schmid, Ann. Phys. (N.Y.) **170**, 333 (1986).

60. C.H. Woo, "Chaos, Ineffectiveness, and the Contrast Between Classical and Quantal Physics," preprint.

61. See Ref. [7] p. 32.

62. D. Sigeti and W. Horsthemke, Phys. Rev. A 35, 2276 (1987).

63. R. Landauer, IEEE Spectrum **4**, 105 (1967).

64. See Ref. [7] p. 3.

65. R. Liboff, private communication.

4
MATRIX COMPUTATIONS AND SIGNAL PROCESSING

CHARLES VAN LOAN
Department of Computer Science
Cornell University
Ithica, NY

4.0 ABSTRACT

The interaction between the signal processing and matrix computation areas is explored by examining some subspace dimension estimation problems that arise in a pair of direction–of–arrival algorithms: MUSIC and ESPRIT. We show that the intelligent handling of these numerical problems requires a successful intermingling of perturbation theory, sensible problem formulation, and reliance upon unitary matrix methods. *

4.1 INTRODUCTION

Signal processing is an application area that has profited by recent research developments in matrix computations. In this paper we examine the synergism between these two fields as suggested by the following diagram:

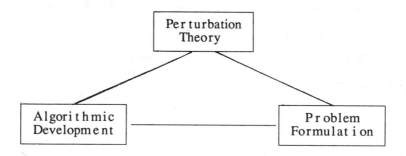

As a vehicle for studying these interactions between computing, mathematics, and engineering we have chosen to examine selected portions of the MUSIC and ESPRIT methods for Direction–of–Arrival (DOA) estimation. The basic problem in DOA estimation is to compute the location of d unknown signals given the output of n sensors. Various assumptions about the signals, the noise, and the array geometry must, of course, be made. We only focus on the interesting linear algebra asssociated with particular implementations of the two algorithms.

---

*This work is partially supported by ONR contract N00014–83–K–640 and NSF contract DCR 86–2310.

In both the MUSIC and ESPRIT procedures the number of signal sources d is estimated by computing the multiplicity of certain eigenvalues. Rank/dimension/multiplicity calculations in the presence of roundoff error and fuzzy data are notoriously tricky and require a good bit of advanced numerical linear algebra. This makes the MUSIC and ESPRIT algorithms ideal for illustrating the more "philosophical" connections betwen matrix computations and signal processing.

The order of presentation is as follows. First, we make some general remarks about perturbation theory and how it can be used in the assessment of a numerical procedure. For the sake of simplicity, we use Gaussian elimination as an example and discuss the importance of the "nearness to singularity" concept. Unitary matrix methods are then shown to be crucial to the intelligent handling of various rank determination problems. The singular value, CS, and Schur decompositions are stressed. After this "trip" into numerical linear algebra we focus on the MUSIC and ESPRIT computations and discuss aspects of their reliable implementation.

Our coverage of these DOA estimation methods is by no means conclusive. There are a few new results but many interesting MUSIC/ESPRIT research questions remain. And again, we are merely using these techniques to dramatize the value of "unitary matrix methodology" in signal processing.

4.2   PERTURBATION THEORY AND ALGORITHM ASSESSMENT

Developing an effective algorithm for a problem and understanding the associated perturbation theory go hand–in–hand in scientific computation. We review various stability/perturbation concepts in the simple setting of linear equation solving. The quality of a linear equation solver cannot be assessed without an understanding of $Ax = b$ sensitivity. How does x change if the elements of A and b are perturbed? An elementary perturbation theory (c.f. Golub and Van Loan (1983, p. 24ff)) tells us that if $(A + \Delta A)(x + \Delta x) = (b + \Delta b)$ and both $\|\Delta A_2\|/\|A\|_2$ and $\|\Delta b\|_2/\|b\|_2$ are $O(\varepsilon)$ with $\varepsilon \ll 1$, then

(H)
$$\frac{\|\Delta x\|_2}{\|x\|_2} \simeq \varepsilon \, \kappa_2(A)$$

Here $\kappa_2(A) = \|A\|_2 \, \|A^{-1}\|_2$ is the 2–norm *condition* of A with respect to inversion. It is easy to show that $\kappa_2(A) > 1$ (always) and that as A gets "close" to being singular, A^{-1} and $\kappa_2(A)$ blow up. The heuristic (H) says that $O(\varepsilon)$ changes in the data A and b usually induce $O(\varepsilon \cdot \kappa_2(A))$ changes in the solution x.

This result can be used to analyze the floating–point performance of Gaussian elimination with pivoting. In particular, if a system $Ax = b$ is solved by this method, then the computed solution $\hat{x}$ exactly solves a "nearby" problem in the sense that

(E) $(A + E)\, \hat{x} = b$ $\|E\|_2 \simeq u\, \|A\|_2$

where u is the machine precision. This shows that the method is stable, i.e., the algorithm does not compound the underlying mathematical sensitivity of the problem. This does <u>not</u> imply that $\hat{x}$ is accurate for if we interpret (E) using (H) then the best thing we can say about $\hat{x}$'s relative error is that $\|\hat{x} - x\|_2 / \|x\|_2 \simeq u\, \kappa_2(A)$. A stable linear equation solver such as Gaussian elimination with pivoting cannot be faulted for producing inaccurate results if the matrix A has a large condition number relative to the machine precision u.

        For many of the basic problems in linear algebra, the perturbation/condition number theory has been worked out and provides a lot of practical guidance in the assessment of algorithms and computed results. See Golub and Van Loan (1983), Stewart (1973), Stewart (1977), and Van Loan (1987).

4.3   RANK DETERMINATION AND THE SINGULAR VALUE DECOMPOSITION

In the linear equation problem a central issue concerns nearness to singularity. More generally we have the problem of estimating the dimension of rectangular matrix range, i.e., rank. From the perspective of pure mathematics, the notion of matrix rank is very crisp:

    If $A \in C^{mxn}$ and $m \geq n$, then A has rank n if and only if it has n independent columns.

    Full column rank is a yes–no, 0–1 proposition. Either a matrix has it or it does not. Unfortunately, fuzzy data and inexact arithmetic complicate the practical treatment of rank. Special tools are needed and one of the most useful in this regard is the singular value decomposition (SVD).

THEOREM 1 (SVD)

    If $A \in C^{mxn}$ $(m \geq n)$ then there exist unitary $U \in C^{mxm}$ and $V \in C^{nxn}$ such that $U^H A V = \Sigma = \mathrm{diag}(\sigma_1 ,..., \sigma_n)$ where $\sigma_1 \geq \sigma_2 \geq ... \geq \sigma_n \geq 0$. The σ_k are called the *singular values*. The columns of U and V are referred to as the corresponding *left and right singular vectors*.

PROOF

    See Golub and Van Loan (1983)

The SVD provides quantitative answers to a number of important questions that arise in practical signal processing work.

Q1.   How close is a matrix to one of lower rank?

A1.   If A has singular values

$$\sigma_1 \geq \dots \sigma_r > \sigma_{r+1} = \dots = \sigma_n = 0$$

then rank(A) = r and

$$\min_{\text{rank}(B) \leq r} \|A - B\|_2 = \sigma_{r+1}.$$

In particular, if $A \in C^{nxn}$ is nonsingular then σ_n is the distance to the set of singular matrices. Thus, $1/\kappa_2(A) = \sigma_n/\sigma_1 = \sigma_n / \|A\|_2$ is a relative measure of nearness to singularity.

Q2.   What is the range and null space of a matrix?

A2.   If $U = [u_1, \dots, u_m]$ and $V = [v_1, \dots, v_n]$ are column partitionings of the left and right singular vector matrices and rank(A) = r then $\text{Null}(A) = \text{span}\{v_{r+1}, \dots, v_n\}$ and $\text{Range}(A) = \text{span}\{u_1, \dots, u_r\}$.

Q3.   How close are two k–dimensional subspaces of C^n ?

A3.   Suppose the n–by–k matrices $Y = [y_1, \dots, y_k]$ and $Z = [z_1, \dots, z_k]$ have orthonormal columns and that $S_1 = \text{Range}(Y)$ and $S_2 = \text{Range}(Z)$. If $\sigma_1 \geq \dots \geq \sigma_k$ are the singular values of $Y^H Z$ then

$$\text{dist}(S_1, S_2) \equiv \min_{\substack{y \in S_1, \ z \in S_2 \\ \|z\|_2 = 1}} \|y - z\|_2 = \sqrt{1 - \sigma_1^2}$$

Q4.   How close are two matrices to having a common null vector?

A4.   If A and B are mxn matrices and $\sigma_1 \geq \ldots \geq \sigma_n$ are the singular values of

$$C = \begin{bmatrix} A \\ B \end{bmatrix}$$

then there exist matrices E_A and E_B satisfying $\|E_A\|_2$, $\|E_B\|_2 \leq \sigma_n$ with the property that

$$\text{Null}(A + E_A) \ \cap \ \text{Null}(B + E_B) \ \neq \ \{\ 0\ \}$$

Proofs of these and other SVD properties can be found in Golub and Van Loan (1983).

4.4 COMPUTING THE SVD

There are a number of ways to compute the SVD. The most important for us is the Golub–Reinsch (1970) procedure which is a derivative of the symmetric QR algorithm. It stably exploits the connection between the SVD of $A \in C^{mxn}$ and the n–by–n Hermitian eigenvalue decomposition $U^H(A^HA)U = \Sigma^H\Sigma$ Implementations may be found in the software packages LINPACK and EISPACK. In either instance the computed $\hat{U}$ and $\hat{V}$ are "unitary" to machine precision u meaning that

$$\|\ \hat{U}^H\hat{U}\ - \ I_m\|_2 \ , \quad \|\ \hat{V}^H\hat{V}\ - \ I_n\|_2 \ \simeq \ u$$

One can also show that the computed singular values are the exact singular values of a matrix $A + \Delta A$ where $\|\Delta A\|_2 \simeq u\|A\|_2$. From this result it is possible to show that the computed singular values $\hat{\sigma}_k$ satisfy

$$|\hat{\sigma}_k - \sigma_k| \ \simeq \ u \cdot \sigma_1 \qquad\qquad k = 1:n$$

This implies that the SVD is guaranteed to detect near rank–deficiency in practice. That is, if A is close to rank deficient then σ_n/σ_1 and its computed analog $\hat{\sigma}_n/\hat{\sigma}_1$ would be small in an order of magnitude sense.

Other methods can be used for SVD computations. If A is sparse then the Lanczos algorithm may be of interest. (See Golub, Luk, and Overton (1981).) In multiprocessor environments the block Jacobi method has exhibited some potential. (See Bischof (1986a, 1986b) and Van Loan (1986).) If just the smallest singular value and associated singular vectors are required, then inverse iteration can be effective. (See Van Loan (1987).)

Throughout this paper we tacitly assume that all SVD computations are performed with the LINPACK implementation of the Golub–Reinsch algorithm.

4.5 UNITARY MATRIX METHODS

The SVD algorithm is but one member of the unitary matrix method family. As we show in subsequent sections, the intelligent handling of numerical rank usually involves use of these methods in conjunction with the SVD.   We summarize them for later reference.

(1)   QR FACTORIZATION

If $A \in C^{m \times n}$ then there exists a unitary $Q \in C^{m \times m}$ such that $Q^H A = R$ is upper triangular.   This is effectively Gram–Schmidt orthogonalization.   If $A = [a_1,...,a_n]$ has rank n and $Q = [q_1,...,q_m]$ then $span\{a_1,...,a_k\} = span\{q_1,...,q_k\}$, $k = 1{:}n$.

(2)   SCHUR DECOMPOSITION

If $A \in C^{n \times n}$ then there exists a unitary $Q \in C^{n \times n}$ such that $Q^H A \, Q = T$ is upper triangular.   The diagonal of T is made up of A's eigenvalues.   If $Q = [q_1,...,q_n]$ then $span\{q_1,...,q_k\}$ is an invariant subspace associated with the eigenvalues $t_{11},...,t_{kk}$.

(3)   HERMITIAN SCHUR DECOMPOSITION

If $A \in C^{n \times n}$ is Hermitian then there exists a unitary $Q \in C^{n \times n}$ such that $Q^H A \, Q = D = diag\,(\lambda_1,...,\lambda_n)$.   If $Q = [q_1,..., q_n]$ is a column partitioning then $Aq_k = \lambda_k q_k$, $k=1{:}n$.

(4)   GENERALIZED SCHUR DECOMPOSITION

If $A,C \in C^{m_1 \times n}$ $(m_1 \geq n)$ and $B,D \in C^{m_2 \times n}$ $(m_2 \geq n)$ then there exist unitary $Q,U,V,$ and Z of appropriate dimension such that the matrices $T_A = U^H A Q$ and $T_B = V^H B Q$ are lower triangular and the matrices $T_C = U^H C Z$ and

$T_D = V^H D Z$ are upper triangular. Since $Q^H (A^H C - \lambda B^H D) Z = T_A^H T_C - \lambda T_B^H T_D$ it follows that this matrix is singular whenever $\lambda = \alpha_i \gamma_i / \beta_i \delta_i$ where the α_i, β_i, γ_i and δ_i are the diagonals of T_A, T_B, T_C, and T_D respectively.

(5)   **CS DECOMPOSITION**

If

$$Q = \begin{bmatrix} Q_1 \\ Q_2 \end{bmatrix} \begin{matrix} m_1 \\ m_2 \end{matrix}$$
$$n$$

satisfies $Q^H Q = I_n$ and m_1 and m_2 are each larger than n, then there exist unitary $U_1 \in C^{m_1 x m_1}$, $U_2 \in C^{m_2 x m_2}$, and $V \in C^{n x n}$ such that

$$\begin{bmatrix} U_1 & 0 \\ 0 & U_2 \end{bmatrix}^H \begin{bmatrix} Q_1 \\ Q_2 \end{bmatrix}^V = \begin{bmatrix} C \\ S \end{bmatrix}$$

where

$$C = diag(\cos(\theta_1) ,..., \cos(\theta_n))$$
$$S = diag(\sin(\theta_1) ,..., \sin(\theta_n))$$

The QR factorization is implemented in LINPACK. The Schur and Hermitian Schur decompositions are part of EISPACK as is the generalized Schur decomposition for the special case $A = B = I_n$, i.e, the problem $Cx = \lambda Dx$.

An algorithm for the generalized Schur decomposition is discussed in Van Loan (1975). The CS decomposition is described in Davis and Kahan (1970), Stewart (1977), Paige and Saunders (1981), Stewart (1983), and Van Loan (1985). Algorithms are given in the last two references. Note that it amounts to a pair of SVDs.

4.6   **THREE EXAMPLES OF PRACTICAL RANK/SUBSPACE DIMENSION DETERMINATION**

The CS and generalized Schur decompositions turn out to be quite important in our discussion of the MUSIC and ESPRIT methods. Before we pursue this, we step through some simpler practical problems that illustrate the value of unitary matrix methods in subspace dimension estimation.

(a)   EIGENVALUE MULTIPLICITY

It is often necessary to deduce the multiplicity of a computed eigenvalue. Suppose A is 2–by–2 and that a unitary matrix Q is found with the property that

$$Q^H(A+E)Q = T = \begin{bmatrix} \lambda_1 & 1 \\ O & \lambda_2 \end{bmatrix} \qquad \|E\| \simeq u\|A\|$$

where u is the machine precision. This says that the computed eigenvalues λ_1 and λ_2 are the exact eigenvalues of a matrix near to A. (The eigenvalue routines in EISPACK permit one to make such statements.) Under what conditions may we assume that the original A has a repeated eigenvalue? A possible criterion is

$$|\lambda_1 - \lambda_2| < u\|A\|$$

However, it is not hard to show that if $|\lambda_1 - \lambda_2| = \varepsilon << 1$, then

$$\sigma_2(A - \lambda_1 I) = O(\varepsilon^2).$$

It then follows from SVD theory that λ_1 is a multiple eigenvalue of a matrix $A + \Delta A$ with $\|\Delta A\|_2 = O(\varepsilon^2)$. Thus, A could be a lot closer to a defective matrix than the mere inspection of the computed eigenvalues reveals. A more reasonable criterion for eigenvalue multiplicity in the n = 2 case would be

$$|\lambda_1 - \lambda_2|^2 < u\|A\|$$

The role of the SVD in eigenvalue multiplicity determination is discussed in Golub and Wilkinson (1976) and Kagstrom and Ruhe (1980).

(b)   RANK DEFICIENT LEAST SQUARES SOLUTION

It can be shown that if $U^H A V = \Sigma$ is the SVD of A with $U = [u_1,...,u_m]$ and $V = [v_1,...,v_n]$, then

$$x_{LS} = \sum_{k=1}^{r} (u_k^H b/\sigma_k)v_k \qquad r = \text{rank}(A)$$

is the minimum 2–norm solution of the least squares problem:

$$\min \|Ax - b\|_2 \qquad A \in C^{mxn}, \; b \in C^m, \; m \geq n$$

The associated minimum residual is then given by

$$\rho_{LS} = \|Ax_{LS} - b\|_2 = \| [u_{r+I} ,...., u_m]^H b \|_2$$

In general, computed singular values are never exactly zero and so we need some procedure for computing an estimate $\hat{r}$ of the rank r. One possibility is to let $\hat{r}$ be the largest integer so that

$$\hat{\sigma}_{\hat{r}} \geq \varepsilon \cdot \hat{\sigma}_1$$

where ε is some small parameter that may depend upon the machine precision and/or the accuracy of the data. With this criterion,

$$\hat{x}_{LS} = \sum_{k=1}^{\hat{r}} (\hat{u}_k^T b / \hat{\sigma}_k) \hat{v}_k$$

can be regarded as a reasonable approximation to the true x_{LS}. Here, the "hat" notation is used to designate computed quantities. Whether or not this is an appropriate way to address a near rank deficient least squares problem depends upon the application. However, with the SVD of A available, the ramifications of an individual $\hat{r}$ choice can be readily explored.

(c)   **SYSTEM CONTROLLABILITY**

Suppose we are given the system

(D) $\hat{x}(t) = Ax(t) + u(t) \cdot b$ $x(0) = x_o$

where $A \in R^{nxn}$, $b \in R^n$, $x_o \in R^n$. Is it possible to find a control function u(t) such that $x(T) = x_T$ where $x_T \in R^n$ is some desired "state" that we wish the system to be in at time T > 0 ? Clearly, this may not be possible. For example, if b = 0 then it is impossible to "control" the x−vector as required. More generally, if the above problem is to have a solution then b must not be deficient in certain directions that can be defined in terms of A's eigensystem. Here are two of the many ways that the "controllability" of (S) can be characterized:

The system (S) is controllable iff $W_1 = [b, Ab, A^2b, ..., A^{n-1}b]$ is nonsingular.

The system (S) is controllable iff $W_2 = \displaystyle\int_0^t e^{At}bb^Te^{A^Tt}\,dt$

is nonsingular.

These are "0–1", yes–no characterizations. Our intuition tells us that if either W_1 or W_2 are nearly singular, then (S) must somehow be "hard" to control. The situation is thus ripe for singular value analysis. The smallest singular value of either W_1 or W_2 may be taken as a measure of how close the system (S) is to being uncontrollable. The use of SVD and related techniques in computational control theory has led to a healthier engineering perspective. Tools now exist for measuring things like nearness to uncontrollability that can assist in the design of robust systems. The spirit of numerical linear algebra's role in control theory is very well illustrated in Paige (1981).

4.7   THE MUSIC PROBLEM

The MUSIC procedure is due to Schmidt (1979, 1981, 1986). We assume that the vector $x \in C^n$ of received waveforms satisfies

$$x = Af + n_x$$

where $A \in C^{nxd}$ is a function of the arrival angles and array element locations, $f \in C^d$ is the vector of incident signals, and $n_x \in C^n$ is the noise. If

$$S_1 = E(xx^H) \qquad\qquad \text{(signal+noise covariance)}$$
$$S_2 = E(n_x n_x^H) \qquad\qquad \text{(noise covariance)}$$
$$P = E(ff^H) \qquad\qquad \text{(incident signal covariance)}$$

then it is easy to see that

$$S_1 = APA^H + S_2$$

The idea behind MUSIC is that under reasonable assumptions we can characterize the number of signals d in terms of the generalized eigenvalues of $S_1 - \lambda S_2$.

THEOREM 2

Let $S_2 \in C^{nxn}$ and $P \in C^{dxd}$ be Hermitian positive definite matrices and assume $S_1 = APA^H + S_2$ where $A \in C^{nxd}$

has rank d.   Then $\lambda = 1$ is the smallest zero of the polynomial $p(\lambda) = \det(S_1 - \lambda S_2)$ and it has multiplicity $n-d$.

PROOF

If $S_1 z = \lambda S_2 z$ with $0 \neq z \in C^n$ then

$$z^H S_1 z = z^H (S_2 + APA^H)z = \lambda \, z^H S_2 z \quad .$$

Thus, $\lambda = 1 + z^H(APA^H)z \, / \, z^H S_2 z$. It follows that $\lambda = 1$ iff $A^H z = 0$. Since $\dim(\text{Null}(A^H A)) = n-d$ it follows that $\lambda = 1$ has multiplicity $n-d$.

After the number of signals d is computed in MUSIC, the DOA's are found by solving a nonlinear minimization problem that requires orthonormal vectors $z_{d+1}, ..., z_n$ such that

(M) $\text{span}\{z_1, ..., z_{n-d}\} = \{ \ x \ | \ S_1 x = \lambda_{min} S_2 x \ \}$

In practice we must estimate the covariance matrices S_1 and S_2. To this end we assume that a "signal–plus–noise" matrix A_S and a "noise alone" matrix A_N have been collected such that

$$A_S{}^H A_S \ \simeq \ E(x \cdot x^H) \ = \ APA^H + E(n_x n_x{}^H) \ \simeq \ APA^H + A_N{}^H A_N$$

The generalized eigenvalue computations now become *generalized singular value computations*.   In particular, we must compute an orthonormal set $\{z_{d+1}, ..., z_n\}$ such that

(M') $\text{span}\{z_{d+1}, ..., z_n\} = \{x \, | \, A_S{}^H A_S z \simeq \lambda_{min} A_N{}^H A_N z\}$

where λ_{min} is the smallest root of $p(\lambda) = \det(A_S{}^H A_S - \lambda A_N{}^H A_N)$. Recognize (M') as an approximation to (M).   The "$\simeq$" is necessary because $A_S{}^H A_S \simeq S_1$ and $A_N{}^H A_N \simeq S_2$ and λ_{min} might not be repeated.   Before we show how to cope successfully with these approximations, it is instructive to examine what may be called an "eigenvector approach" to (M).

4.8   MUSIC AND THE GENERALIZED SINGULAR VALUE PROBLEM

Generalized singular value problems of the form

$$A_S{}^H A_S x \ = \ \lambda A_N{}^H A_N x$$

have a number of nice properties that we state without proof.

THEOREM 3

If $A_S \in C^{m_1 \times n}$ $(m_1 \geq n)$ and $A_N \in C^{m_2 \times n}$ $(m_2 \geq n)$ then there exists a nonsingular $X \in C^{n \times n}$ such that

$$X^H(A_S^H A_S)X = \text{diag}(\alpha_1 ,..., \alpha_n) = D_S \qquad \alpha_k \geq 0$$

and

$$X^H(A_N^H A_N)X = \text{diag}(\beta_1 ,..., \beta_n) = D_S \qquad \beta_k \geq 0$$

PROOF

See Golub and Van Loan   (1983, p.314)

The quantities $\sqrt{\alpha_k / \beta_k}$ are called the generalized singular values of the pair $\{A_S, A_N\}$ and the columns of X the associated generalized singular vectors. Note that if $X = [x_1 ,..., x_n]$ is a column paritioning, then

$$\beta_k(A_S^H A_S)x_k - \alpha_k(A_N^H A_N)x_k = 0 \qquad\qquad k=1:n$$

Let's assume that all the β_k are nonzero. It follows that if $\lambda_k = (\alpha_k/\beta_k)^2$ then $(A_S^H A_S - \lambda_k A_N^H A_N)x = 0$. One approach to (M') would be to proceed as follows:

(a)   Find X and order the columns so that the $\lambda_k = \alpha_k/\beta_k$ range from large to small.

(b)   Determine d such that $\lambda_d > \lambda_n + \epsilon \geq \lambda_{d+1} \cdots \geq \lambda_n \geq 0$ for some small tolerance ϵ.

(c)   Compute the QR decomposition $ZR = [x_{d+1} ,..., x_n]$. The columns of Z form the desired orthonormal basis.

The "standard" methods for computing X are all flawed and are instructive to look at. Again we assume that A has full rank which guarantees nonzero B.

METHOD 1

1. $S_1 = A_S{}^H A_S$; $S_2 \leftarrow A_N{}^H A_N$ /* Form cross–products:

2. $S_2 = LL^H$ /* L = Cholesky lower triangle

3. Solve: $LW = S_1$; $LC = W^H$ /* $C = L^{-1} S_1 L^{-H}$

4. $Q^H CQ = D_S = \mathrm{diag}(\alpha_k)$ /* Compute C's eigensystem, $Q^H Q = 1$

5. Solve: $L^H X = Q$ /* $D_N = I_n$

This method is of dubious quality in that if A_N is nearly rank deficient then C will be highly contaminated with error. This makes it impossible to compute the small eigenvalues accurately. These are precisely the eigenvalues of interest in MUSIC.

METHOD 2

1. $S_1 = A_S{}^H A_S$; $S_2 \leftarrow A_N{}^H A_N$ /*Form cross–products:

2. $S_2 = U_2 D U_2{}^H$, $D = \mathrm{diag}(\lambda_1)$ /* Hermitian Schur decomposition

3. $Y \leftarrow UD^{-1/2}$; $C \leftarrow Y^H S_1 Y$ /* $Y^H S_2 Y = I$

4. $Q^H CQ = \mathrm{diag}(\alpha_1 ,..., \alpha_n)$ /* Compute C's eigensystem, $Q^H Q = 1$

5. $X = YQ$;   /* $D_N = I_n$

This is similar to Method 1 except that ill–conditioning in A_N is immediately identified because the eigensystem of S_2 is obtained. C will again be very poorly determined if S_2 is nearly rank deficient.

METHOD 3

1. $U_2{}^H A_N V_2 = \mathrm{diag}(\sigma_k)$ /* SVD

2. $Y \leftarrow V_2 \, \mathrm{diag}(1/\sigma_1 ,..., 1/\sigma_n)$

3. $A_S \leftarrow A_S Y$ /* $Y^H S_2 Y = I$

4. $U_1^H A_S V_1 = D$ /* SVD

5. $X \leftarrow Y V_1$ /* $D_S = D^H D$; $D_N = I_n$

The only difference between this and Methods 1 and 2 is that the cross products $A_S^H A_S$ and $A_N^H A_N$ are avoided. It is well–known that a significant loss of information can occur in numerical cross products. That is why the method of normal equations is often unsatisfactory when solving least square problems. That is why the singular values of a matrix A are <u>not</u> obtained by computing the eigenvalues of $A^H A$. See Golub and Van Loan (1983, pp. 143–289).

The point in presenting these three methods is to see how vulnerable the MUSIC computations are to rank deficiency. Unfortunately, near rank deficiency in A_N is not unusual. It is sometimes possible to circumvent this by interchanging the roles of A_S and A_N in the above and hoping that the columns of A_S are strongly independent. Instead of pursuing this line we present a method that doesn't care so much about near rank deficiency in the data.

4.9   A COMPLETELY UNITARY APPROACH TO MUSIC

The following approach to problem (M') is proposed in Speiser and Van Loan (1984).

METHOD 4

1. Compute the QR factorization

$$\begin{bmatrix} A_S \\ A_N \end{bmatrix} = \begin{bmatrix} Q_1 \\ Q_2 \end{bmatrix} R$$

where Q_1 and Q_2 have the same size as A_S and A_N respectively and $R \in C^{n \times n}$ is upper triangular. Assume that R is nonsingular, i.e., that Null(A_S) ∩ Null(A_N) = { 0 }.

2. Compute the CS Decomposition

$$\begin{bmatrix} Q_1 \\ Q_2 \end{bmatrix} = \begin{bmatrix} U_1 & 0 \\ 0 & U_2 \end{bmatrix} \begin{bmatrix} C \\ S \end{bmatrix} V$$

where U_1, U_2, and V are unitary, $C = \text{diag}(\cos(\theta_k))$, and $S = \text{diag}(\sin(\theta_k))$ with $0 \leq \theta_1 \leq \cdots \leq \theta_n \leq \pi/2$. It follows that if $X = R^{-1}V$ then $X^H(A_S^H A_S - \mu^2 A_N^H A_N)X = C^H C - \lambda S^H S$ and so the generalized singular values are specified by $\mu_k = \text{ctn}(\theta_k)$.

3. Define $\hat{d}$ by $c_{\hat{d}} > \varepsilon + c_n \geq c_{\hat{d}+1} \geq \cdots c_n \geq 0$ where $\varepsilon > 0$ is a small positive tolerance. Here, $c_k \equiv \cos(\theta_k)$.

4. Compute the QR factorization of the product $ZT = R^H V$ where $Z = [z_1, ..., z_n]$ is unitary and $T \in C^{n \times n}$ is upper triangular. Since

$X = R^{-1}V = (V^H R)^{-1} = ((R^H V)^H)^{-1} = ((ZT)^H)^{-1} = ZT^{-H}$ and T^{-H} is lower triangular, it follows that $\text{span}\{z_{\hat{d}+1}, ..., z_n\} = \text{span}\{x_{\hat{d}+1}, ..., x_n\}$.

Note that both $\hat{d}$ and the basis $\{z_{\hat{d}+1}, ..., z_n\}$ are found without any inversions or cross–products, the computations that undermine the reliability of Methods 1–3. Moreover, Method 4 is not prone to the sensitivity of the eigenvectors $\{x_{\hat{d}+1}, ..., x_n\}$. It is possible for an eigenspace to be well–conditioned even though the eigenvectors that define it are not. This is a common theme in many applications that require a basis for an eigenspace. Orthonormal bases are usually preferable to eigenvector bases from the numerical point of view. This is why the Schur decomposition is preferable to the Jordan decomposition when doing invariant subspace computation.

We conclude with a result about what Method 4 actually computes in light of the tolerance that is used to define $\hat{d}$.

THEOREM 4

The vectors $\{z_{\hat{d}+1}, ..., z_n\}$ produced by Method 4 exactly span the minimum singular value subspace for a problem $\tilde{A}_S^H \tilde{A}_S - \lambda \tilde{A}_N^H \tilde{A}_N$ where

$$\left\| \begin{bmatrix} \tilde{A}_S \\ \tilde{A}_N \end{bmatrix} - \begin{bmatrix} A_S \\ A_N \end{bmatrix} \right\|_2 \leq \varepsilon \, \|R\|_2$$

PROOF

Define $\tilde{C} = \text{diag}(\cos(\hat{\tilde{\theta}}_k))$ and $S = \text{diag}(\sin(\hat{\tilde{\theta}}_k))$ where $\tilde{\theta}_k = \theta_k$ if $k \leq \hat{d}$ and $\tilde{\theta}_k = \theta_k$ is $k > \hat{d}$. Set $\tilde{A}_S = U_1 \tilde{C} V^H R$ and $\tilde{A}_N = U_2 \tilde{S} V^H R$. Since

$$\tilde{A}_S - A_S = U_1(\tilde{C} - C)V^H R$$

and

$$\tilde{A}_N - A_N = U_2(\tilde{S} - S)V^H R$$

it follows that

$$\left\| \begin{bmatrix} \tilde{A}_S \\ \tilde{A}_N \end{bmatrix} - \begin{bmatrix} A_S \\ A_N \end{bmatrix} \right\|_2 = \left\| \begin{bmatrix} (\tilde{C} - C) \\ (\tilde{S} - S) \end{bmatrix} V^H R \right\|_2 \leq \epsilon \|R\|_2$$

Thus, if MUSIC is implemented using Method 4 it solves a "nearby problem" exactly.

4.10 THE ESPRIT PROBLEM

ESPRIT avoids the nonlinear minimization in MUSIC. It does this by comparing output x and y from a pair of sensor arrays (X and Y), one a translate of the other. Details may be found in Paulraj, Roy, and Kailath (1986) and Roy, Paulraj, and Kailath (1986). A comparison of the MUSIC and ESPRIT procedures can be found in Roy, Paulraj, and Kailath (1987).

In ESPRIT the output of the two arrays is modelled as follows:

$$x(t) = A\,s(t) + n_x(t)$$

$$y(t) = A\Phi s(t) + n_y(t)$$

where (it can be shown), $A \in C^{nxd}$, n_x and n_y are the noise vectors, $s(t) \in C^{dxd}$ is the vector of source signals, and Φ is a diagonal unitary matrix whose diagonal entries are easy functions of the DOAs. Under certain assumptions we have

$$S_{XX} = E(xx^H) = ASA^H + \sigma^2 I$$

$$S_{YY} = E(yy^H) = A\Phi S\Phi^H A^H + \sigma^2 I$$

$$S_{XY} = E(xy^H) = AS\Phi^H A^H$$

In ESPRIT, the computation of Φ is based upon the following result.

THEOREM 5

Suppose $A \in C^{n \times d}$ has rank d, $S \in C^{d \times d}$ is Hermitian positive definite, and $\Phi \in C^{d \times d}$. If $S_{XX} = ASA^H + \sigma^2 I$ and $S_{YY} + A\Phi S\Phi^H A^H + \Phi^2 I$ then Φ^2 is their smallest eigenvalue and in either case it has multiplicity n–d. Likewise if $V = [v_1, ..., v_n]$ is unitary with

$$S_{XX} v_k = \Phi^2 v_k \qquad\qquad k = d+1:n$$

then $A^H V = [B^H \; O]$ where $B \in C^{d \times d}$ and thus,

$$V^H S_{XX} V = \begin{bmatrix} BSB^H + \sigma^2 I_d & 0 \\ 0 & \sigma^2 I_{n-d} \end{bmatrix}$$

and

$$V^H S_{YY} V = \begin{bmatrix} B\Phi S\Phi^H B^H + \sigma^2 I_d & 0 \\ 0 & \sigma^2 I_{n-d} \end{bmatrix}$$

If $S_{XY} = AS\Phi^H A^H$ then

$$V^H S_{XY} V = \begin{bmatrix} BS\Phi^H B^H & 0 \\ 0 & 0 \end{bmatrix}$$

There are d complex numbers $\lambda_1, ..., \lambda_d$ for which

$$\text{rank}((S_{XX} - \sigma^2 I) - \lambda S_{xy}) = d-1$$

and these are precisely the λ that make $BSB^H - \lambda BS\Phi^H B^H = BS(I - \lambda \Phi^H)B^H$ singular.

PROOF

If $S_{XX}v = \lambda v$ with $\|v\|_2 = 1$, then $\lambda = v^H(ASA^H + \sigma^2 I)v$
$= \sigma^2 + (A^H v)^H S(A^H v)$. It follows that σ^2 is the smallest eigenvalue of S_{XX} and that it has multiplicity $n-d =$ dim(Null(A^H)). If $S_{XX}v = \sigma^2 v$ then $v \in$ Null(A^H). Thus, if $V = [v_1,...,v_n] \in C^{n \times n}$ is unitary with span$\{v_{d+1},...,v_n\} =$ Null(A^H), then $A^H V$ has the form specified in the hypothesis. The rest of the proof is straight forward.

In ESPRIT, Φ is diagonal and so we may take $\Phi = \text{diag}(\lambda_1,....,\lambda_d)$ where $\lambda = \lambda_k$ forces rank $((S_{XX} - \sigma^2 I) - \lambda S_{XY}) = d-1$

If the matrices S_{XX}, S_{YY}, and S_{XY} are known exactly then there are several ways to compute Φ:

METHOD E1

1. Compute the Hermitian Schur decomposition of S_{XX} : $V^H S_{XX} V = D$.
Since $\lambda = \sigma^2$ is an eigenvalue of multiplicity $n-d$ we may assume

$$V^H S_{XX} V = \left[\begin{array}{cc} \text{diag}(\sigma_k^2) & 0 \\ 0 & \sigma^2 I \end{array} \right] \begin{array}{c} d \\ n-d \end{array}$$
$$\quad\quad\quad\quad\quad d \quad\quad n-d$$

If $V = [v_1,...,v_n]$ it follows that span$\{v_{d+1},...,v_n\} =$ Null(A^H).

2. Compute $W = V^H S_{XY} V$ which must have the form

$$V^H S_{XX} V = \left[\begin{array}{cc} W_{11} & 0 \\ 0 & 0 \end{array} \right] \begin{array}{c} d \\ n-d \end{array}$$
$$\quad\quad\quad\quad\quad d \quad\quad n-d$$

3. Compute $\lambda_1,...\lambda_d$ such that $\det(\text{diag}(\sigma_k^2 - \sigma^2) - \lambda_k W_{11}) = 0$ for k=1:d and set $\Phi = \text{diag}(\lambda_1,...,\lambda_d)$.

METHOD E2

1. Compute the Schur decomposition of S_{XY} : $Q^H S_{XY} Q = T$.　Since this matrix has a null space of dimension $n{-}d$ we can chose Q such that

$$Q^H S_{XY} Q = \begin{bmatrix} T_{11} & 0 \\ 0 & 0 \end{bmatrix} \begin{matrix} d \\ n{-}d \end{matrix}$$
$$\phantom{Q^H S_{XY} Q = }\;\; d \quad\; n{-}d$$

Note that if $Q = [q_1,...,q_n]$ then $\mathrm{span}\{q_{d+1},...,q_n\} = \mathrm{Null}(A^H)$.

2. Compute $Q^H S_{XX} Q = W$ which must have the form

$$Q^H S_{XX} Q = \begin{bmatrix} W_{11} & 0 \\ 0 & \sigma^2 I \end{bmatrix} \begin{matrix} d \\ n{-}d \end{matrix}$$
$$\phantom{Q^H S_{XX} Q = }\;\; d \quad\; n{-}d$$

3. Compute $\lambda_1,...,\lambda_d$ such that $\det((W_{11} - \sigma^2 I) - \lambda_k T_{11}) = 0$, $k{=}1{:}d$. Set $\Phi = \mathrm{diag}(\lambda_1,...,\lambda_d)$

METHOD E3

　　　　Essentially the same as Method E1, but with S_{YY} replacing S_{XX}.

Thus, the approach taken in all these methods is to factor out the common null space $\mathrm{Null}(A^H)$ and solve the remaining $d{-}by{-}d$ generalized eigenproblem.

　　　　In practice, a great deal of care must be exercised because we will only have approximations to S_{XX}, S_{YY} and S_{XY}.　Indeed, if we collect m snapshots and form the matrices

$$A_X = (1/\sqrt{m}\,) [x(t_1),...,x(t_m)\,]^T \in C^{mxn}$$

then

$$A_Y = (1/\sqrt{m}\,) [y(t_1),...,y(t_m)\,]^T \in C^{mxn}$$

then $A_X{}^H A_X \simeq S_{XX}$, $A_Y{}^H A_Y \simeq S_{YY}$, $A_X{}^H A_Y \simeq S_{XY}$.　The trouble now is that we must now factor an approximate nullspace out of the approximate problem

$$(A_X{}^H A_X - \sigma^2 I) - \lambda A_X{}^H A_Y$$

Generalized eigenvalue problems of the form $A - \lambda B$ where A and B have nearly intersecting nullspaces (the ESPRIT situation) are notoriously ill–conditioned and great care must be exercised.

4.11 A SINGULAR VALUE APPROACH

This corresponds to Method E1. Note that if

$$U_X^H A_X V_X \;=\; \Sigma_X \;=\; \mathrm{diag}(\sigma_{k,X})$$

is the SVD of A_X and

$$\sigma_{\hat{d},X} \;>\; \varepsilon + \sigma_{n,X} \;\geq\; \sigma_{\hat{d}+1,X} \;\geq\; \cdots \;\geq\; \sigma_{n,X} \;=\; \sigma_{min,X}$$

where ε is a small tolerance, then the corresponding columns of V_X are an approximate basis for $\mathrm{Null}(A^H)$:

$$\mathrm{span}\{v_{\hat{d}+1,X},...,v_{n,X}\} \;\simeq\; \mathrm{Null}(A^H) \quad .$$

This is because $A_X^H A_X \simeq S_{XX} = ASA^H + \sigma_2 I$. Likewise, if

$$U_Y^H A_Y V_Y \;=\; \Sigma_Y \;=\; \mathrm{diag}(\sigma_{k,Y})$$

is the SVD of A_Y and

$$\sigma_{\hat{d},Y} \;>\; \varepsilon + \sigma_{n,Y} \;\geq\; \sigma_{\hat{d}+1,Y} \;\geq\; \cdots \;\geq\; \sigma_{n,Y}$$

then

$$\mathrm{span}\{v_{\hat{d}+1,Y},...,v_{n,Y}\} \;\simeq\; \mathrm{Null}(A^H) \quad .$$

Unfortunately, these two singular vector approximations to $\mathrm{Null}(A^H)$ may differ, even in their choice of d.   A way around this difficulty that assigns "equal weight" to the X and Y data is to compute the intersections of the subspaces

$$\mathrm{span}\{v_{\hat{d}+1,X},...,v_{n,X}\} \;\cap\; \mathrm{span}\{v_{\hat{d}+1,Y},...,v_{n,Y}\}$$

for various values of $\hat{d}$ and settle on that value that gives the largest possible dimension. Subspace intersections can be computed using the SVD, see Golub and Van Loan (1983,p.430.). Moreover, if we have

$$V_X^H V_Y \;=\; V \;=\; \begin{bmatrix} V_{11} & V_{12} \\ V_{21} & V_{22} \end{bmatrix} \begin{matrix} \hat{d} \\[4pt] n-\hat{d} \end{matrix}$$
$$\qquad\qquad\qquad\qquad \hat{d} \qquad n-\hat{d}$$

and $\|V_{12}\|_F = \delta$, then $\text{span}\{v_{\hat{d}+1,X},...,v_{n,X}\}$ and $\text{span}\{v_{\hat{d}+1,Y},...,v_{n,Y}\}$ are "within δ" of being the same subspace. This follows from CS decomposition theory. Here, $\|\cdot\|_F$ denotes the sum–of–squares norm. Note how easy it is to compute $\|V_{12}\|_F$ as a function of d. This could be used in an intelligent way to compute the critical basis for $\text{Null}(A^H)$.

Once we have unitary $V = [v_1,...,v_n]$ with $\text{span}\{v_{\hat{d}+1},...,v_n\} \simeq$ $\text{Null}(A^H)$ and as an estimate $\hat{\sigma}^2$ of σ^2, say $\hat{\sigma}^2 = \sigma_{n,X}{}^2$, then our problem transforms as follows

$$V^H \left[(A_X^H U_X U_X^H A_X - \hat{\sigma}^2 I) - \lambda A_X^H U_X U_X^H A_Y\right] V \;=\;$$

$$\begin{bmatrix} B^H B - \sigma^2 I & E_{12} \\ E_{21} & E_{22} \end{bmatrix} - \lambda \begin{bmatrix} B^H C & F_{12} \\ F_{21} & F_{22} \end{bmatrix}$$

Here, B and C are the upper $\hat{d}$–by–$\hat{d}$ portions of $U_X^H A_X V$ and $U_Y^H A_Y V$ respectively and the E_{ij} and F_{ij} are small in norm. How small depends upon the quality of the approximations $\text{span}\{v_{d+1},...v_n\}$. There remains the problem of solving the d–by–d eigenproblem

$$\det[(B^H B - \sigma^2 I) - \lambda B^H C] \;=\; \det[(\tilde{B} + \sigma I)(\tilde{B} - \sigma I) - \lambda B^H C] \;=\; 0$$

where $\tilde{B}$ is Hermitian matrix that satisfies $\tilde{B}^2 = B^H B$. This can be found from the SVD $B = U_B \Sigma_B V_B^H$: Just set $\tilde{B} = V_B \Sigma_B V_B^H$.

To avoid cross products and inverses, the above determinantal equation could be solved using the technique in Van Loan (1975) that computes the generalized Schur decomposition described earlier. In general, the computed λ will not be on the unit circle as they should be in theory. Thus, it is plausible to set $\Phi = \text{diag}(\text{arg}(\lambda_k))$.

Although this procedure relies on unitary matrices throughout, we have been unable to show that this implementation of ESPRIT computes the exact DOAs of a "nearby" problem.

4.12   A GENERALIZED SCHUR APPROACH

We next outline a unitary matrix approach to the ESPRIT problem that corresponds to Method E2.

1. Compute unitary Q and V such that

$$V^H A_X^H Q = T$$
$$Q^H A_Y V = S$$

are upper triangular. Note that $V^H A_X^H A_Y V = TS = R \in C^{nxn}$ is the Schur decomposition of $A_X^H A_Y$ and so V can be chosen to order the eigenvalues from largest to smallest in absolute value.

It is possible to do this without forming $A_X^H A_Y$ using the algorithm in Van Loan (1975). Determine $\hat{d}$ so that

$$|r_{\hat{d}\hat{d}}| = |s_{\hat{d}\hat{d}} t_{\hat{d}\hat{d}}| > |r_{nn}| + \varepsilon = |s_{nn} t_{nn}| + \varepsilon \geq |s_{\hat{d}+1,\hat{d}+1} \cdot t_{\hat{d}+1,\hat{d}+1}|$$

where ε is a small tolerance. Note that if $V = [v_1,... \ v_n]$ then $\text{span}\{v_{\hat{d}+1},...,v_n\} \simeq \text{Null}(A^H)$.

2. If $U^H(A_X V)$ is upper triangular, then it has the form

$$U^H(A_X V) \simeq \begin{bmatrix} W_{11} & 0 \\ 0 & \sigma^2 I \end{bmatrix} \begin{matrix} \hat{d} \\ n-\hat{d} \end{matrix}$$

3. Compute the generalized eigenvalues of the problem

$$(\tilde{W}_{11} + \sigma I)(\tilde{W}_{11} - \sigma I) - \lambda \ S(1:d,1:d) \cdot T(1:d,1:d)$$

using the algorithm in Van Loan (1975) and set $\Phi = \text{diag}(\text{arg}(\lambda_k))$.

Here W_{11} is a Hermitian matrix that satisfies $\tilde{W}_{11}^2 = W_{11}^H W_{11}$.

As with the singular value approach, we are not able to show that this implementation of EPRIT solves a nearby problem. Thus, the stability properties of ESPRIT are unclear to us although good unitary methods exist for the computations.

4.13   CONCLUSIONS

We have shown how the difficult subspace dimension estimation problems in MUSIC and ESPRIT can be handled. In the case of MUSIC, we are able to show that the computed DOAs are exact for "nearby data". This shows that the method is stable.

ESPRIT is conceptually much simpler, but involves a trickier eigenvalue computation. More research is necessary to examine how the ESPRIT DOAs are effected by the choice of d and the computed basis for $Null(A^H)$. In the mean time, simulations suggest that ESPRIT is pretty reliable, prompting us to conjecture that there is a favourable perturbation analysis of the method that awaits discovery.

4.14   REFERENCES

1.   C. Bischof (1986a), "A parallel ordering for the block Jacobi method on a hypercube architecture", Cornell Computer Science Technical Report TR 86–740.

2.   C. Bischof (1986b), "Computing the singular value decomposition on a distributed system of vector processors", Cornell Computer Science Report TR 86–798.

3.   R. Brent, F. Luk, and C. Van Loan (1985), "Computing the singular value decomposition using mesh–connected processors", *J. of VLSI and Computer Systems*, 1, 242–270.

4.   C. Davis and W.Kahan (1970), "The rotation of eigenvectors by a perturbation III", *SIAM J. NUMER. ANAL.*, 7,1–46.

5.   J. Dongarra, C.B. Moler, J.R. Bunch, and G.W. Stewart (1979), *LINPACK User's Guide*, SIAM Publications, Philadelphia.

6.   G.H. Golub and C. Reinsch (1970), "Singular value decomposition and least squares", *NUMER. MATH.*, 14, 403–420.

7.   G.H. Golub, F.T. Luk, and M. Overton (1981), "A block Lanczos method for computing the singular values and corresponding singular vectors of a matrix", *ACM TRANS. MATH. SOFT.*, 7, 149–169.

8.   G.H. Golub and C. Van Loan (1983),   MATRIX COMPUTATIONS, Johns Hopkins University Press, Baltimore, Md.

9.   G.H. Golub and J. H. Wilkinson (1976),   "Ill–conditioned eigensystems and the computation of the Jordan Canonical Form", *SIAM Review*, 18, 578–619.

10.   B. Kagstrom and A. Ruhe (1980), "An algorithm for the numerical computation of the Jordan normal form of a complex matrix", *ACM. Transactions on Mathematical Software*, 6, 398–419.

11.   C.C. Paige (1981), "Properties of Numerical algorithms related to computing controllability", *IEEE Trans. Auto. Cont.*, AC–26, 130–138.

12.   C.C. Paige and M.A. Saunders (1981), "Towards a generalized singular value decomposition", *SIAM J. NUMER. ANAL.*, 18, 398–405.

13.   A. Paulraj, R. Roy, and T. Kailath (1986a), "A subspace rotation signal parameter estimation", *Proc. of the IEEE*, 1044–1045.

14.   R.Roy, A. Paulraj, and T. Kailath (1986), "ESPRIT– a subspace rotation approach to estimation of cisoids in noise", *IEEE Trans. Acoustics, Speech and Signal Processing*, ASSP–34(4).

15.   R. Roy, A. Paulraj, and T. Kailath (1987), "Comparative performance of ESPRIT and MUSIC for Direction–of–Arrival Estimation", Proc., Int'l Conference on Acoustics, Speech and Signal Processing, Dallas, TX., 2344–2347.

16.   R.O. Schmidt (1979), "Multiple emitter location and signal parameter estimation", Proc. RADC Spectrum Estimation Workshop, Griffiths AFB, NY.

17.   R.O. Schmidt (1981), *A signal subspace approach to multiple emitter location and spectral estimation*, Ph.D. dissertation, Dept. of Electrical Engineering,Stanford.

18.   R.O. Schmidt (1986), "Multiple emitter location and signal parameter estimation", *IEEE Trans. on Antennas and Propagation*, AP–34, 276–280.

19.   J. Speiser and C. Van Loan (1984), "Signal processing computations using the generalized singular value decomposition", Proc. of SPIE, Vol. 495, SPIE International Symposium, San Diego, August, 1984.

20.   G.W. Stewart (1973), "Error and perturbation bounds for subspaces associated with certain eigenvalue problems", *SIAM Review*, 15, 727–764.

21.   G.W. Stewart (1977), "On perturbation of pseudo–inverses, projections, linear least squares problems." *SIAM Review*, 19, 634–662.

22.   G.W. Stewart (1983), "An algorithm for computing the CS for a partitioned orthonormal matrix", *NUMER. MATH.*, 40, 297–306.

23.   C. Van Loan (1975), "A general matrix eigenvalue problem", *SIAM J. NUMER. ANAL.*, 12, 819–834.

24.   C. Van Loan (1976), "Generalizing the singular value decomposition", *SIAM J. NUMER. ANAL.*, 13, 76–83.

25.   C. Van Loan (1984), "Analysis of some Matrix Problems using the CS Decomposition", Cornell Computer Science Technical Report TR 84–603.

26.   C. Van Loan (1985), "Computing the CS and generalized singular value decompositions", *NUMERISCHE MATHEMATIK*, 46, 479–491.

27.   C. Van Loan (1986), "Computing the singular value decomposition on a ring of array processors", in *LARGE SCALE EIGENVALUE PROBLEMS*, J. Cullum and R. Willoughby (eds), Elsevier, 51–66.

28.   C. Van Loan (1987), "On estimating the condition of eigenvalues and eigenvectors", *Lin. Alg. & Applic.*, 88/89, 715–732.

5

SYSTOLIC LINEAR ALGEBRA MACHINES: A SURVEY

ROBERT SCHREIBER
Computer Science Department
Rensselaer Polytechnic Institute
Troy, New York

5.1 INTRODUCTION

Matrix computation is coming to play an important role in signal processing, especially in direction finding and beamforming for sensor arrays. A number of new algorithms employ matrix singlular value and eigenvalue decompositions [72]. The standard optimal method in beamforming requires the solution of a least squares problem [97].

These matrix computations all require $O(mn^2)$ arithmetic operations for an $m \times n$ matrix. In a typical application, the matrix is formed by taking m samples or "snapshots" of the output of an n–element sensor array. Thus, the matrix can be acquired in $O(m)$ time. Clearly, real–time solution of these problems means $O(m)$–time solution, and this requires a parallel process with $O(n^2)$ elements. Fortunately, systolic arrays having $O(n^2)$ processing elements are known for the necessary matrix computations.

This paper is a personal view of the history of systolic linear algebra. Although similar ideas already existed, Kung and Leiserson, in 1978, made the importance of systolic arrays in VLSI computer design widely known [47]. Their first designs were elegant, efficient hardware realizations of the obvious algorithms for some basic matrix problems: banded matrix multiply, solving a banded triangular system of equations, and banded LU decomposition by Gaussian elimination without pivoting. It was quite clear from the outset that the systolic technique could be used to build, out of highly replicated VLSI cells, dedicated, special purpose devices for kernel algorithms that would have extraordinary performance, whether measured in arithmetic operations per dollar–second, per kilogram–second, or per watt–second (joule).

Since then, intensive research has been devoted to a number of very important questions that arose naturally from the promising idea of systolic arrays:

1.   (The algorithmic question). What problems have algorithms that can be realized as systolic arrays? If the standard algorithm is not systolic, can a better algorithm be found?

2.   (The hardware systems question). How do we build systolic systems? What VLSI devices should be used? How do we supply the array with data and control? How much memory per cell? How much prgrammability? Should the array be synchronous or not? How to deal with hardware faults?

73

3.   (The theoretical question).   Can we automatically design a systolic array, given a procedural description of an algorithm? Can we define the class of programs for which a systolic realization exists?

4.   (The application question).   What difficult tasks (especially in signal processing) can we successfully address with systolic arrays.

Great progress has been made on these fronts since 1978.   Concerning the theoretical question, refer to Rao's Ph.D thesis [70], which contains an excellent survey of the literature. Rao has largely solved the problem, although it is not clear whether his approach covers all relevant systolic arrays and systolic algorithms. Subsequent work that extends Rao's methods has been done by Delosme and Ipsen [25].   Precursors of Rao's thesis were papers by Cappello and Stieglitz [21], Moldovan [61] and Quinton [69].   The thesis of Lisper [51] is a thorough analysis of the problem of synthesis of arbitrarily connected synchronous hardware systems including but not limited to systolic arrays.   Moldovan and Fortes [62] have begun an analysis of the problem of automatically decomposing large matrix problems for solution on arrays of fixed size.

For the hardware question, the work of Kung, et. al., [2,26,35,36,48], and of several industrial groups [6,19,46,63,64,93] has been very important. There are now two very powerful, programmable, systolic machines in existence, the 100 Mflop Warp [2,26,41], built by Kung's group at CMU, and the 1000 Mflop Saxpy Matrix−1 [37,84], built by Saxpy Computer Corporation.

Huang and Abraham [42] and Luk [56] have pioneered an approach to fault tolerance that uses row and column check sums and is able to detect and correct certain errors in matrix computations a posteriori and at very low cost. Used by the host machine, these techniques could allow for some systolic arrays to fail gracefully. It is not clear how far this technique can be extended. Whether it applies to the eigenvalue problem is an open question.

Speiser and Whitehouse [91,97] have been instrumental in highlighting the potential importance of systolic arrays in signal processing, particularly as a tool for implementation of Schmidt's MUSIC algorithm [72] and other high−resolution methods for direction finding and beamforming.   The Chimera project at ESL, Inc. has successfully demonstrated the immense value of systolic hardware for real−time sonar signal processing [66].

There has been a plethora of articles on systolic implementation of (among other things) queues [50], systolic sorting, pattern matching, integer and polynomial greatest common divisor computation [11,12], combinatorial and graph algorithms [40], and dynamic programming.

5.2   SYSTOLIC LINEAR ALGEBRA

Kung and Leiserson's first arrays were wonderfully intriguing, but did not address any really important linear algebraic computations. Since then, many significant and useful arrays have been proposed.   As is customary in discussions of matrix computation, we will divide the field as follows:

1. Linear Systems

2. Least Squares Problems

3. Eigenvalue Problems

4. Singular Value Decomposition

5.2.1 LINEAR SYSTEMS

The problem is to solve $Ax=b$ where A is a given $n \times n$, nonsingular matrix and b is a given $n \times 1$ vector.

This problem is largely solved. At first, there seemed to be a real difficulty: the standard library routines employ Gaussian elimination with partial pivoting, an algorithm that is not sufficiently systolic. In any systolic implementation of partial pivoting, only $O(n)$ processors can be used and $O(n^2)$ time is required. But there are several alternative methods that have linear–time systolic implementations. These are

1. Neighbor pivoting. This algorithm, a variant of Gaussian elimination, was long used, but discarded in favor of partial pivoting because of the latter's more solid stability theory. Gentleman and Kung introduced it for systolic arrays [38], and Sorensen has recently analyzed its stability [90]. In practice it behaves quite well. Both dense [38] and banded [78] matrices can be handled.

2. QR decomposition. The algorithm is this: given A,b, one computes an upper triangular matrix $R=Q^tA$ and a vector $\tilde{b}=Q^tb$ where the orthogonal matrix Q is a product of plane rotations. Then one solves the triangular system $Rx=b$. This is always stable, but has three times the operation count of Gaussian elimination. The systolic arrays are due to Bojanczk, Brent and Kung (square dense matrices) [7], Gentleman and Kung (rectangular matrices) [38], and Heller and Ipsen (banded matrices) [41]. Generalizations [78] and implementation issues [77] were discussed by Schreiber and Kuekes. A clever, systolic, Jacobi–like method for QR decomposition of a square matrix was later invented by Luk [53].

The Gentleman–Kung array is shown in Figure 1. In this array, the circular cells generate plane rotations that annihilate the elements of the input matrix A below the diagonal. The right–pointing connections propagate the rotations through the array, where they are applied to the remainder of the matrix and to the right–hand–side vector b by the square cells. The upper triangular factor R and the vector Q^tb reside in the cells of the array. There are about $n^2/2$ cells in the array and the computing time is n clocks.

The Heller–Ipsen array is shown in Figure 2. This array takes advantage of bandedness in the matrix to reduce the number of cells: for a matrix with p nonzero subdiagonals and q nonzero superdiagonals, an array of size $p \times (p+q+1)$ is used.

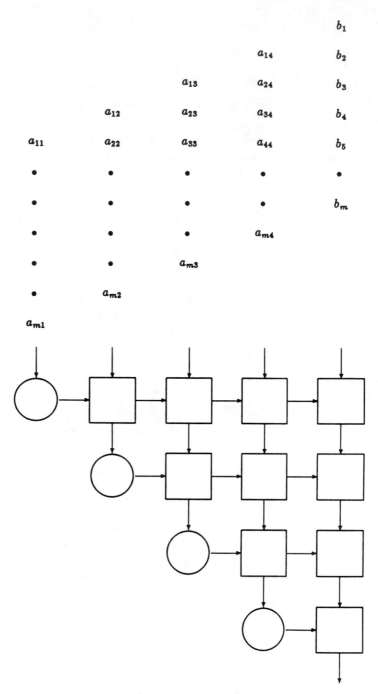

Figure 1   The Gentleman–Kung Array.

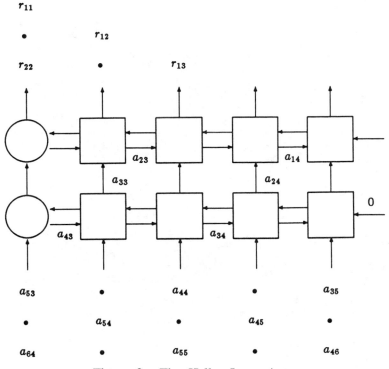

Figure 2   The Heller–Ipsen Array

Schreiber and Kuekes pointed out how a large matrix can be decomposed by several passes of the data through an undersized array of the structure given in Figure 3.

3.   A number of papers point out the advantages of some less–used methods for linear systems in the systolic context. For example, Ahmed, Delosme, and Morf [1] and Delosme and Ipsen [25] show that symmetric positive definite systems can be solved in one pass through an array that uses a nonstandard algorithm based on hyperbolic plane rotations.   Delosme and Ipsen have extended this work to the case of Toeplitz systems [24].   Robert and Tchuente have developed other one–pass arrays for general systems of linear equations [71].

4.   Schulz' iterative algorithm [87] may be used, since this requires only matrix multiplication.   The algorithm was analyzed extensively by Soderstrom and Stewart [89], who found that is not practical because of its operation count.   Schreiber [73] suggested several improvements that may make it a practical alternative in some special contexts.

An important question is whether the systolic notion applies only to dense matrices.   Melhem has attempted to extend it to arbitrary sparse

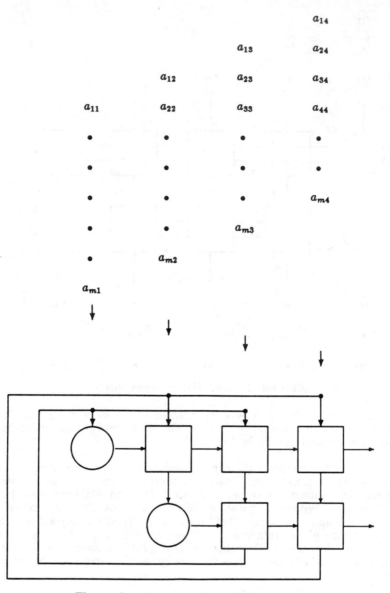

Figure 3   The Schreiber–Kuekes Array.

matrices [60].   Worley and Schreiber show how to achieve linear–time solution to partial difference equations on $n \times n$ grids using the nested dissection method [98]; straight forward use of a systolic band solver for this problem would require $O(n^2)$ time.

Systems with special matrices call for special arrays. Toeplitz systems can be solved in linear time with a linear array [14, 49]. This can also be done for Vandermonde systems. The Cholesky factorization of a symmetric positive definite band matrix requires half as many processors as a nonsymmetric band matrix [13]. This result can be extended to full matrices [85].

Another important problem is updating factorizations after a low–rank change to the matrix. Schreiber and Tang discuss a variety of updates of symmetric positive definite matrices and the Cholesky and LDL^T factorization [79]. New algorithms for updating least–squares solutions and eigendecompositions (and related signal processing problems) after obtaining some additional signal vectors are given by Karasalo [45] and by Schreiber [80].

Fadeeva's method [33] is attractive since it can be implemented by a single array, while the usual decomposition and backsolve methods require two arrays.

To interface different systolic arrays, the output pattern of one must be matched with the input pattern to the next. Storing intermediated data in memories is a general but costly and potentially slow solution. A systolic method for transposing a matrix, which would be used in such a situation, is described by O'Leary [65].

5.2.2 LEAST SQUARES PROBLEMS

The QR decomposition is the preferred tool [39]. Systolic QR is due to Bojanczyk, Brent, and Kung (only square matrices) [7], Gentleman and Kung (rectangular matrices) [38], and Heller and Ipsen (banded matrices) [41]. Elaboration and extension is due to Schreiber and Kuekes [77, 78] and McWhirter, et. al. [59,96]. Barlow and Ipsen discuss the use of square–root free methods [3,44]. Methods based on hyperbolic rotations are given by Ahmed, Delosme and Morf [1]. For least squares linear prediction (in which the matrix is rectangular and Toeplitz) see Bojanczyk [10].

Often, ill–conditioned least–squares problems have to be solved. In this case, it is useful to know the rank of the data matrix. To get some information about the rank of a matrix from the QR decomposition, column pivoting is often used [39]. But this does not reliably indicate rank. Moreover, column pivoting cannot readily be incorporated into a systolic QR method (without increasing the time complexity from $O(n)$ to $O(n^2)$). Fortunately, there is a good alternative. Chan has recently shown that the rank of a matrix may be more reliably estimated through a standard QR factorization and some relatively inexpensive post–processing of R [22].

Regularization (ridge regression) is an important technique for handling ill–conditioned least squares problems [39]; and these often arise in signal processing. Elden has given efficient algorithms for solving regularized least–squares problems by QR factorization of the structured, block 2×1 matrices that arise, both for the general [29] and the banded [30] case. These methods are implemented systolically by Elden and Schreiber (general and banded cases) [31] (the array is depicted in Figure 4) and by Bojanczyk and Brent [9] and Elden and Schreiber [32] (Toeplitz case).

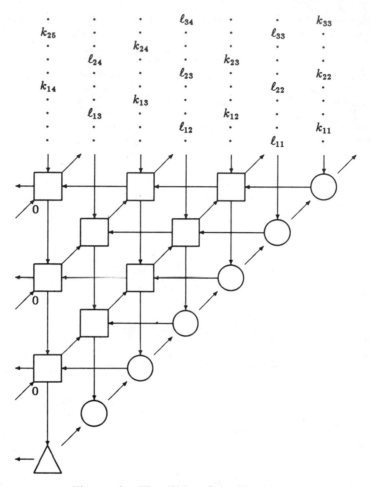

Figure 4   The Elden–Schreiber Array.

5.2.2.1   MATRIX MULTIPLY AS A KERNEL

Bischof and Van Loan [5] have recently pointed out that algorithms that are "rich in matrix multiply" are now very attractive, largely because matrix multiply can be so easily done at extra–high speed by a systolic array. They described a partitioning of the Householder QR factorization for which most of the work is in the form of matrix multiply.   In fact, if a partitioning of the matrix into k block columns is used, then $(k-1)/k$ is the fraction of the work accounted for by matrix multiplication, while only $1/k$ of the work is in the nonmatrix multiply residue.   Schreiber and Parlett [83] have suggested a true block QR method (which is an extension of an earlier version due to Dietrich [27]) that is more complex, but for which the residue is only $O(k^{-2})$.

Schulz's method, mentioned above, is another algorithm of interest because of its richness in matrix multiply.   When applied to a rectangular matrix, possibly not of fullrank, Schulz's algorithm produces the generalized inverse, as shown by Ben–Israel and Cohen [4]. This allows it to be used to produce minimum norm solutions of rank deficient least squares problems [39]. Schreiber's modification improves the robustness of the method for the generalized inverse [73].

5.2.3 EIGENVALUE PROBLEMS

The nonsymmetric eigenvalue problem is, in general, quite difficult to solve since it is often very ill–conditioned. The algorithm of choice is to reduce the matrix to upper Hessenberg form and then use the QR iteration. No one has shown how to accommmplish this in less than $O(n^2)$ time using systolic arrays. Stewart has described a systolic implementation of a Jacobi–like method that requires linear time per sweep. But the convergence is very slow unless the matrix is normal [92]. Eberlein has recently demonstrated a more elaborate Jacobi method that seems to be more robust than Stewart's [28]. Fortunately, the eigenvalue problems arising in signal processing are symmetric (or complex Hermitian).

For the problem of finding all the eigenvalues and eigenvectors of a symmetric matrix, a near linear time solution using an $n/2 \times n/2$ array was described by Brent and Luk [15,16]. Their first paper [15] is one of the landmarks in systolic matrix computation. Until then, much fruitless effort had been devoted to looking for fast implementations of the QR algorithm [76]. Brent and Luk realized that the Jacobi method, coupled with a new parallel sweep order that they had invented, would be more appropriate for systolic arrays. Furthermore, they discovered that their parallel order was more effective than the usual cyclic–by–rows order, even on a sequential machine. This is an empirical result, obtained by experiments on large numbers of matrices with random elements. Recently, two proofs of convergence of several systolic Jacobi algorithms have appeared. Schwiegelshohn and Thiele [86] and Luk [57,58] have obtained similar proofs.

Schreiber has given an efficient block Jacobi strategy to solve the eigenvalue problem with a Brent/Luk array even when the matrix is too large for the array [75]. Another approach to this problem was given by Scott, Heath, and Ward [88]. Block methods are a general strategy for partitioning large matrices for computation by several processors in parallel, or by several passes through a fixed systolic array. See [82] for an overview of block methods. Further analysis of and implementation issues in block Jacobi methods were given by Van Loan [95].

Although its mapping into systolic arrays is very clean, the Jacobi method does have two drawbacks: it is considerably slower than QR, and it does not take advantage of bandedness in the matrix. Therefore, interest in other methods continues. Recently, two papers have shown how to tridiagonalize a matrix in $O(n \log n)$ time: Bojanczyk and Brent using a square array [8] and Schreiber, using a more complex array structure that more naturally takes advantage of bandedness [81].

The question then is how to solve the tridiagonal eigenvalue problem. Two alternatives are known: to use bisection [73] which takes linear time, or to use Cuppen's method [23] with a systolic implementation of the eigenvalue updating procedure due to Bunch, Nielsen, and Sorensen [20], which is at the heart of Cuppen's method.

It is clear, however, that any systolic implementation of these methods will be considerably more complicated than the Brent/Luk Jacobi array.

5.2.4   SINGULAR VALUE DECOMPOSITION

Early work on systolic methods for the singular value decomposition (SVD) by Finn, Luk, and Pottle [34], Ipsen [43], Schreiber [74], and Brent and Luk [16] was surpassed by Brent, Luk, and Van Loan [17] who gave an $O(n\log n)$ Jacobi–like method. The method, due to Kogbetliantz, involves the repeated solution of 2×2 SVD problems. Paige has recently proved the ultimate quadratic convergence of Kogbetliantz's method [67].

As in the case of the eigenvalue problem, there appear to be competitive methods based on bidiagonalization and solving the bidiagonal problem [81].

The generalized SVD, which is of use in solving constrained least squares problems [94] and in the analysis of signal subspace problems with nonwhite noise, can be computed with similar techniques [18,55,68]. The related and increasingly important CS decomposition can be computed by a sequence of systolic SVD computations as described by Luk [54].

Before applying the iteration method described by Brent, Luk, and Van Loan, it is necessary first to orthogonally reduce the matrix to a square matrix. For this, a QR factrization may be used. Luk has given some interesting ways to combine these two steps in one array [52].

5.3   THE SAXPY MATRIX–1

Because it is costly to develop special–purpose hardware, whether at the chip, board, or system level, it has become useful to develop general–purpose, programmable, systolic systems. These are programmable computers that have integrated processor arrays, memory, and a host computer. They perform especially well on systolic algorithms.

We shall briefly describe one such machine, the Saxpy Matrix–1. More detail than we can give here is provided in [37]. The system architecture is shown in Figure 5.

The System Controller is a general–purpose computer that runs the application program and which interfaces to low–speed I/O devices. It is a VAX and runs the VMS operating system. The application program makes calls to "primitive" subroutines that run in the systolic Matrix Processor.

The System Memory is where data arrays reside. Its capacity is 128 Mwords. It is implemented with low–cost, 256 Kbit, MOS dynamic RAM chips. Through 32–way interleaving it provides 320 million bytes per second of bandwidth.

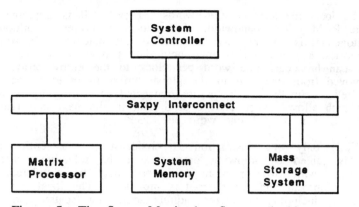

Figure 5   The Saxpy Matrix–1:   System Architecture.

System memory is dynamically allocated.   The memory manager (a software module) maintains the size, shape, and location of arrays in memory. When an operation is performed by the systolic Matrix Processor, its operands are sections (i.e., subarrays) of arrays in system memory.

The Matrix Processor architecture is shown in Figure 6.   It is an array of 32 cells, each of which performs floating–point computation on data in its local memory and on data that is broadcast to all the cells.   The array is synchronous, single–instruction, multiple–data (SIMD):   thus, every cell performs the same operation on every clock, but on different data.   Since there are 32 multiplies and 32 adds performed every 64 nsec clock, the peak speed is 1000 million floating–point operations per second.

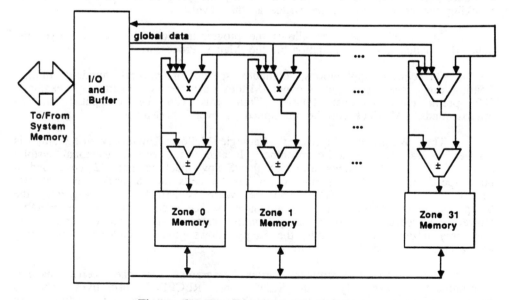

Figure 6   The Saxpy Matrix Processor.

The local memory is 4096 words per cell. It is implemented in fast, ECL static RAM. The bandwidth of the local memories, 4 billion words per second altogether, is one key to the speed of the machine. In addition to the local memories, a global buffer may hold data that is to be shared by all the cells. It can broadcast one word per clock to the entire array. Also, data can be moved from one cell to the global buffer in order to be broadcast to all other cells. Finally, the cells are connected in a ring by a systolic path. This data path allows every cell to send data to the next cell (and from Cell 31 to Cell 0) at a rate of one word per clock.

While only one instruction is broadcast to the entire array, some flexibility is gained by allowing processors to be selectively disabled. Thus, systolic arrays with two or more cell types may be simulated. Hardware support for reciprocal and square root is provided. The local memory supports vector accesses including vectorized indirect addressing (scatter and gather).

Data arrays are moved in blocks from the System Memory to the local memories over the Saxpy Interconnect at 250 million bytes per second. the loading and storing of data blocks can be overlapped with computation in the Matrix Processor.

The Matrix Processor is programmed in subsets of FORTRAN–77 and FORTRAN–8x. The compiler detects and exploits systolic parallelism in the code. It also handles blocking of computation; moving subarrays of arrays into the local memory of the Matrix Processor, performing a computation, and returning the data to the system memory.

A second "medium–level" language is FORTRAN–8x like in syntax, but distinguishes between system–memory resident variables, local memory variables, and global buffer variables. Thus, the programmer is responsible for blocking the computation if he writes at this level.

An assembly language allows the programmer complete control over the detailed scheduling of computation in the Matrix Processor.

On typical applications, compute speeds of several hundred million floating–point operations per second (Mflops) can be sustained. For example, 1024–point complex Fast Fourier Transforms can be computed in 75 microseconds. Matrices can be multiplied at over 700 Mflops.

The Warp, developed at Carnegie–Mellon University [2,26,48] is another programmable systolic machine. It also has a one–dimensional systolic array. The chief differences are that the Saxpy machine has 32 cells and a peak speed of 1000 Mflops, the Warp has 10 cells and a peak speed of 100 Mflops. The Saxpy is SIMD and the Warp is MIMD. Consequently the Warp is more flexible, but communications between cells is considerably slower. Moreover, the Warp array communicates to the system memory and host only through the end cells; the Saxpy allows global broadcast to the array and access to any cell of the array.

The Warp programmer writes programs for the cells, handles communications explicitly with SEND and RECEIVE primitives, and is responsible for decomposing the algorithm to run on the given number of cells; the Saxpy programmer may do these things, but may also ignore these issues and allow the compiler to do it for him. On the other hand, the Saxpy programmer is responsible for deciding what is done in the host and what is done in the array, while the Warp compiler automatically generates code for both array and host.

5.4 CONCLUSIONS

We end by noting some important unresolved questions.

* What is the best systolic method for the nonsymmetric eigenvalue problem? Must any linear–time method be sensitive to non–normality?

* We need a full explanation of automatic methods for partitioning big problems for solution on small arrays.

* We need good high–level programming languages and tools for using the programmable arrays that have been and are now being developed.

5.5 ACKNOWLEDGEMENTS

I would like to thank Professor Charles Van Loan for his helpful comments on this paper.

Partial support by the Office of Naval Research under Contract N00014–86–K–0610 and by the Army Research Office under Contract DAAL03086–K–0112 is gratefully acknowledged.

5.6 REFERENCES

1.   H.M. Ahmed, J.M. Delosme, and M. Morf, Highly concurrent structures for matrix arithmetic and signal processing, IEEE Computer 15 (1982) 65–82.

2.   M. Annaratone, E. Arnould, T. Gross, H.T. Kung, M. Lam, O. Menzilcioglu, K. Sarocky, and J.A. Webb, Warp architecture and implementation, Proceedings of the 13th Annual International Symposium on Computer Architecture, IEEE 1986, pp. 346–356.

3.   J.L. Barlow and I.C.F. Ipsen, Parallel scaled givens rotations for the solution of linear least squares problems, Report DCS/RR–310, Dept. of Computer Sci., Yale University, 1984.

4.   A. Ben–Israel and D. Cohen, On iterative computation of generalized inverses and associated projections, SIAM Journal on Numerical Numerical Analysis 3 (1966) 410–419.

5.   Christian Bischof and Charles Van Loan, The WY representation for products of Householder matrices, SIAM Journal on Scientific and Statistical Computing 8 (1987) s2–s13.

6.   J. Blackmer, G. Frank, and P. Kuekes, A 200 million operations per second (MOPS) systolic processor, In Tian F. Tao, editor, Real–Time Processing IV, pp. 10–18, SPIE vol. 298. SPIE, Bellingham, WA, 1982.

7.   A Bojanczyk, R. P. Brent, and H. T. Kung, Numerically stable solution of dense systems of linear equations using mesh–connected processors Journal on Scientific and Statistical Computing 5 (1984) 95–104.

8.   A. Bojanczyk and R.P. Brent, Tridiagonalization of a symmetric on a square array of mesh–connected processors, Journal of Parallel and Distributed Computing 2 (1985) 261–276.

9.   A. Bojanczyk and R. P. Brent, Parallel solution of certain Toeplitz least squares problems, Lincar Algebra and Its Applications 77 (1986) 43–60.

10.   A. W. Bojaczyk, Systolic implementation of the lattice algorithm for least squares linear prediction problems, Linear Algebra and Its Applications 77 (1986) 27–42.

11.   R.P. Brent and H.T. Kung, Systolic VLSI arrays for linear–time GCD computation, In F. Anceau and E. J. Aas, editors, VLSI '83, pp. 145–154, North–Holland, August, 1983.

12.   R.P. Brent and H.T. Kung, A systolic algorithm for integer GCD computation, Report TR–CS–82–11, Dept. of Computer Science, The Australian National University, 1982 (revised 1984).

13.   Richard P. Brent and Franklin T. Luk, Computing the Cholesky factorization using a systolic architecture, Proceedings of the 6th Australian Computer Science Conference, Australian Computer Science Communications 5:1 (1983) 295–302.

14.   Richard P. Brent and Franklin T. Luk, A systolic array for the linear–time solution of Toeplitz systems of equations, Dept. of Computer Science, Cornell University, TR–82–526, 1982.

15.   R.P. Brent and F.T. Luk, A systolic architecture for almost linear time solution of the symmetric eigenvalue problem, Department of Computer Science, Cornell University, TR–82–525, August 1982.

16.   R.P. Brent and F.T. Luk, The solution of singular value and symmetric eigenvalue problems on multiprocessor arrays, SIAM Journal on Scientific and Statistical Computing 6 (1985) 69–84.

17. R.P. Brent, F.T. Luk, and C. Van Loan, Computation of the singular value decomposition using mesh–connected processors, Journal of VLSI Computer Systems 1 (1985) 242–270.

18. R.P. Brent, F.T. Luk, and C. Van Loan, Computation of the generalized singular value decomposition using mesh–connected processors, SPIE Volume 341, Real Time Signal Processing VI, pp. 66–71, SPIE, Bellingham, WA, 1983.

19. K. Bromley, J.J. Symanski, J.M. Speiser, and H.J. Whitehouse, Systolic array processor developments, Technical Report, Naval Ocean Systems Center, October, 1981.

20. J.R. Bunch, C.P. Nielsen, and D.C. Sorensen, Rank–one modification of the symmetric eigenproblem, Numerische Mathematik 31 (1978) 31–48

21. P.R. Capello and K. Stieglitz, Unifying VLSI Array Designs and Geometric Transformations, Proceedings of the 1983 International Conference on Parallel Processing, IEEE, 1983.

22. Tony F.C. Chan, Rank Revealing QR factorization, Linear Algebra and Its Applications 88 (1987) 67–82.

23. J.J.M. Cuppen, A divide and conquer method for the symmetric tridiagonal eigenproblem, Numerische Mathematik 36 (1981) 177–195.

24. Jean–Marc Delosme and Ilse Ipsen, Efficient Systolic Arrays for the Solution of Toeplitz Systems: An illustration of a Methodology for the Construction of Systolic Architectures in VLSI, International Symposium on VLSI Technology, Systems and Applications, pp. 268–273, May 1985.

25. Jean–Marc Delosme and Ilse Ipsen, Parallel solution of symmetric positive definite systems with hyperbolic rotations, Linear Algebra and Its Applications 77 (1986) 75–111.

26. Department of Computer Science, Carnegie Mellon University, Pittsburgh, PA 15213, Collection of Papers on Warp.

27. G. Dietrich, A new formulation of the hypermatrix Householder–QR decomposition, Computer Methods in Applied Mechanics and Engineering 9 (1976) 273–280.

28. P. J. Eberlein, On the Schur decomposition of a matrix for parallel computation, IEEE Transactions on Computers C–36 (1987) 167–174.

29. Lars Elden, Algorithms for the regularization of ill–conditioned least squares problems, BIT 17 (1977) 134–145.

30. Lars Elden, An algorithm for the regularization of ill–conditioned, banded least squares problems, SIAM Journal on Scientific and Statistical Computing 5 (1984) 229–236.

31.   L. Elden and R. Schreiber, An application of systolic arrays to linear, discrete ill–posed problems, SIAM Journal on Scientific and Statistical Computing 7 (1986) 892–904.

32.   L. Elden and R. Schreiber, A systolic array for the regularization of ill–conditioned least–squares problems with triangular Toeplitz matrix, Linear Algebra and Its Applications 77 (1986) 137–147.

33.   D.K. Fadeev and V.N. Fadeeva, Computational Methods of Linear Algebra, W. H. Freeman, San Francisco, 1963.

34.   Alan M. Finn, Franklin T. Luk and Christopher Pottle, Systolic array computation of the singular value decomposition, SPIE vol. 341, Real Time Signal Processing V, pp. 35–43, SPIE Bellingham, WA, 1982.

35.   A.L. Fisher, H.T. Kung, L.M. Monier, H. Walker and Y. Dohi, Design of the PSC: a programmable systolic chip, In R. Bryant, editor, Proceedings of the Third Caltech Conference on Very Large Scale Integration, pp. 287–302. California Institute of Technology, Computer Science Press, Inc., March 1983.

36.   A.L. Fisher, H.T. Kung, L.M. Monier and Y. Dohi, Architecture of the PSC: a programmable systolic chip, Proceedings of the 10th Annual International Symposium on Computer Architecture, pp. 48–53, June 1983.

37.   David Foulser and Robert Schreiber, The Saxpy Matrix–1: A general–purpose systolic computer, IEEE Computer 20:7 (1987).

38.   W.M. Gentleman and H.T. Kung, Matrix triangularization by systolic arrays, SPIE vol. 298, Real–time Signal Processing IV, pp. 19–26. SPIE, Bellingham, WA, 1981.

39.   Gene H. Golub and Charles F. Van Loan, Matrix Computation, Johns Hopkins, Baltimore, Maryland, 1983.

40.   L.J. Guibas, H.T. Kung and C.D. Thompson, Direct VLSI implement– ation of combinatorial algorithms, Proceedings of Conference on Very Large Scale Integration: Architecture, Design, Fabrication, pp. 509–525, California Institute of Technology, January 1979.

41.   D.E. Heller and I.C.F. Ipsen, Systolic networks for orthogonal decompositions, SIAM Journal on Scientific and Statistical Computing 4 (1983) 261–269.

42.   K.H.K. Huang and J.A. Abraham, Low cost schemes for fault tolerance in matrix operations with processor arrays, Technical Report, University of Illinois, 1982.

43.   I.C.F. Ipsen, Singular value computatations with systolic arrays, SPIE vol. 549, Real Time Signal Processing VII, pp. 13–21. SPIE, Bellingham, WA, 1984.

44.   I.C.F. Ipsen, A parallel QR method using fast Givens rotations, Report
        DCS RR–299, Dept. of Computer Sci., Yale University, 1984.

45.   Ilkke Karasalo, Estimating the covariance matrix by signal subspace
        averaging, IEEE Transactions on Acoustics, Speech, and Signal
        Processing ASSP–34 (1986) 8–12.

46.   A.V. Kulkarni and D.Y. Yen, Systolic processing and an
        implementation for signal and image processing, IEEE Transactions on
        Computers C–31 (1982) 1000–1009.

47.   H.T. Kung and C.E. Leiserson, Systolic arrays (for VLSI), In I.S.
        Duffand G.W. Stewart, editors, Sparse Matrix Proceedings 1978, pp.
        256–282. Society for Industrial and Applied Mathematics, 1979. A
        slightly different version appears in Introduction to VLSI Systems by
        C.A. Mead and L.A. Conway, Addison–Wesley, 1980, Section 8.3.

48.   H.T. Kung, Architecture of the Warp processor, In Keith Bromley,
        editor, Critical Review of Technology: Highly Parallel Signal
        Processing, SPIE vol. 601, SPIE, Bellingham, WA, 1986.

49.   S.Y. Kung and Y.H. Hu, A highly concurrent algorithm and pipelined
        architecture for solving Toeplitz systems, IEEE Transactions on
        Acoustics, Speech, and Signal Processing, ASSP–31 (1983) 66–76.

50.   Charles E. Leiserson, Systolic priority queues, Technical Report,
        Carnegie–Mellon University, 1979.

51.   Bjorn Lisper, Synthesizing Synchronous Systems by Statis Scheduling in
        Space–time, Ph.D. thesis, The Royal Institute of Technology, Stockholm,
        Sweden, 1987.

52.   Franklin T. Luk, A triangular processor array for computing singular
        values, Linear Algebra and Its Applications 77 (1986) 259–273.

53.   Franklin T. Luk, A rotation method for computing the QR–
        decomposition, SIAM Journal on Scientific and Statistical Computing 7
        (1986) 452–459.

54.   Franklin T. Luk and Sanzheng Aiao, Computing the CS–decomposition
        on systolic arrays, SIAM Journal on Scientific and Statistical
        Computing 8 (1987) 1121–1125.

55.   Franklin T. Luk, A parallel method for computing the generalized
        singular value decomposition, Journal of Parallel and Distributed
        Computing 2 (1985) 250–260.

56.   Franklin T. Luk, Algorithm–based fault tolerance for parallel matrix
        equation solvers, SPIE Vol. 564, REalTime Signal Processing VIII, pp.
        49–53, SPIE, Bellingham, WA, 1985.

57.   Franklin T. Luk, On parallel Jacobi orderings, Report EE–CEG–86–5,
        Cornell University, June 1986.

58.   Franklin T. Luk, A proof of convergence for two parallel Jacobi SVD algorithms, IEEE Transactions on Computers, to appear.

59.   J.G. McWhirter, Systolic array for recursive least–squares minimisation Electronics Letter 19(18):729–730, September 1983.

60.   Rami Melhem, Parallel solution of linear systems with striped sparse matrices, Part 2: Stiffness Matrices, A Case Study, Parallel Computing, to appear.

61.   Dan I. Moldovan, On the design of algorithms for VLSI systolic arrays, Proceedings of the IEEE 71 (1983) 113–120.

62.   Dan I. Moldovan and Jose A.B. Fortes, Partitioning and mapping algorithms into fixed size systolic arrays, IEEE Transactions on Computers C–36 (1986) 1–12.

63.   J. Greg Nash, A systolic/cellular compputer architecture for linear algebraic operations, 1985 IEEE International Conference on Robotics and Automation, pp. 779–784, 1985.

64.   J. Greg Nash and C. Petrozolin, VLSI implementation of a linear systolic array, Proceedings ICASSP 85, IEEE, 1985, pp. 1392–1395.

65.   Diane P. O'Leary, Systolic arrays for matrix transpose and other reorderings, IEEE Transactions on Computers C–36 (1987) 117–122.

66.   N. Owsley, Systolic array adaptive beamforming, Chapter 6 of these Proceedings.

67.   C.C. Paige and P. Van Dooren, On the quadratic convergence of Kogbetliantz's algorithm for computing the singular value decomposition, Linear Algebra and Its Applications 77 (1986) 301–313.

68.   C.C. Paige, Computing the generalized singular value decomposition SIAM Journal on Scientific and Statistical Computing 8 (1987) 1126–1146.

69.   P. Quinton, Automatic synthesis of systolic arrays from uniform recurrent equations, Proceedings of the 11th Annual International Symposium on Computer Architecture, IEEE, 1984, pp. 208–214.

70.   Sailesh Rao, Regular Iterative Algorithms and Their Implementation on Parallel Processor Arrays, Ph.D. Thesis, Stanford University, 1985.

71.   Yves Robert and Maurice Tchuente, Resolution systolique de systemes lineaires, Mathematical Modeling and Numerical Analysis 19 (1985) 315–326.

72.   Ralph Otto Schmidt, A Signal Subspace Approach to Multiple Emitter Location and Spectral Estimation, Ph.D. Thesis, Stanford University, 1981.

73.   Robert Schreiber, Computing generalized inverses and the eigenvalues of symmetric matrices using systolic arrays, In R. Glowinski and J.L. Lions, editors, Computing Methods in Applied Sciences and Engineering, North–Holland, pp. 285–295, 1984.

74.   R. Schreiber, A Systolic architecture for singular value decomposition Proceedings 1er Colloque International sur les Methodes Vectorielles et Paralleles en Calcul Scientifique, Electicite de France Bulletin de la Direction des Etudes et Recherches, Serie C, No. 1 (1983) 143–148.

75.   Robert Schreiber, Computing the eigenvalues of a symmetric matrix on an undersized systolic array, SIAM Journal on Scientific and Statistical Computing 7 (1986) 441–451.

76.   Robert Schreiber, Systolic arrays for eigenvalue computation, SPIE vol. 341, Real Time Signal Processing V, paper 341–50, SPIE Bellingham, WA, 1982.

77.   R. Schreiber and P.J. Kuekes, Systolic linear algebra machines in digital signal processing, In S.Y. Kung, J.J. Whitehouse and T. Kailath, editors, VLSI and Modern Signal Processing, Prentice–Hall, 1985, pp. 389–405.

78.   Robert Schreiber, On systolic array methods for band matrix factorizations BIT 26 (1986) 303–316.

79.   Robert Schreiber and Wei Pai Tang, On Systolic arrays for updating the Cholesky factorization BIT 26 (1986) 451–466.

80.   Robert Schreiber, Implementation of adaptive array algorithms, IEEE Transactions on Acoustics, Speech, and Signal Processing ASSP–34 (1986) 1038–1045.

81.   Robert Schreiber, Tridiagonalization and symmetric bidiagonalization by systolic array, Research Report, Saxpy Computer Corporation, Sunnyvale, CA, 1985. Submitted to IEEE Transactions on Computers.

82.   Robert Schreiber, Block methods for parallel machines, Report 87–5, Dept. of Computer Science, Rensselaer Polytechnic Institute, Troy, NY 12180, 1987.

83.   Robert Schreiber and Beresford Parlett, Block reflectors: theory and computation, SIAM Journal on Numerical Analysis, to appear.

84.   Robert Schreiber, The Saxpy–1M: architecture and algorithms, In Keith Bromley, editor, Critical Review of Technology: Highly Parallel Signal Processing, SPIE vol. 601, SPIE, Bellingham, WA, 1986.

85.   Robert Schreiber, Cholesky factorization by systolic array, Technical Report 87–14, Dept. of Computer Science, Rensselaer Polytechnic Institute, Troy, New York. Submitted to IEEE Transactions on Computers.

86. U.Schwiegelshohn and L.Thiele, A systolic algorithm for cyclic–by–rows SVD, Proceedings of the IEEE International Conference on Acoustics Speech and Signal Processing, 1987.

87. G. Schulz, Iterative berechnung der reziproken Matrix, Z. Angew, Math. Mech. 13 (1933) 57–59.

88. David S. Scott, Michael T. Heath and Robert C. Ward, Parallel block Jacobi eigenvalue algorithms using systolic arrays, Linear Algebra and Its Applications 77 (1986) 345–355.

89. Torsten Soderstrom and G.W. Stewart, On the numerical properties of an iterative method for computing the Moore–Penrose generalized inverse, SIAM Journal on Numerical Analysis 11 (1974) 61–74.

90. Danny Sorensen, Analysis of pairwise pivoting in Gaussian elimination Report ANL/MCS–TM–26, Argonne National Labs, 1984.

91. J.M. Speiser and H.J. Whitehouse, Parallel processing algorithms and architectures for real–time signal processing, SPIE vol. 298, Real Time Signal Processing IV, pp. 2–9, SPIE, Bellingham, WA, 1982.

92. G.W. Stewart, A Jacobi–like algorithm for computing the Schur decomposition of a non–Hermitian matrix, SIAM Journal on Scientific and Statistical Computing 6 (1985) 853–864.

93. J.J. Symanski, A systolic array processor implementation, Technical Report, Naval Ocean Systems Centre, 1981.

94. Charles Van Loan, On the method of weighting for equality constrained least squares problems, SIAM Journal on Numerical Analysis 22 (1985) 851–864.

95. Charles Van Loan, The block Jacobi method for computing the singular value decomposition, In C. Byrnes and A. Ludquist, editors, Proceedings of the Seventh International Symposium on the Mathematical Theory of Networks and Systems. Also available as Report TR 85–680, Dept. of Computer Sci., Cornell University, 1985.

96. C.R. Ward, A.J. Robson, P.J. Hargrave and J.G. McWhirter, The application of a systolic least square processing array to adaptive beamforming, Proceedings of ICASSP 84, pp. 34A.3.1–34A.3.4. Standard Telecommunication Labs and Royal Signals and Radar Establishment, 1984.

97. H.J. Whitehouse, J.M. Speiser and K. Bromley, Signal Processing applications of concurrent array processor technology, In S.Y. Kung, H.J. Whitehouse and T. Kailath, editors, VLSI and Modern Signal Processing, Prentice–Hall, 1985, pp. 25–41.

98. Patrick H. Worley and Robert Schreiber, Nested dissection on a mesh–connected processor array, In Arthur Wouk, editor, New Computing Environments: Parallel, Vector, Systolic, SIAM, Philadelphia, 1986, pp. 8–38.

6
SYSTOLIC ARRAY ADAPTIVE BEAMFORMING

NORMAN L. OWSLEY
Underwater Systems Centre
Newport, Rhode Island

6.1   INTRODUCTION

Optimum algorithms for the space–time processing requirements of a discrete sensor array have existed in one form or another for almost twenty–five years [1].   The computational requirements for implementation have inhibited the widespread application of these techniques to broadband arrays.   To date, the hardware realization of the intensive linear algebra operations required for any type of modern signal processing function has not been amenable to cost–effective solutions.   In this regard, the most promising new development in modern signal processor design has been the result of rapid progress in very large scale integration (VLSI) technology.   Large scale integration has allowed the fabrication of special purpose components, which has provided the impetus for the evolution of very powerful special–purpose computing architectures for linear algebra intensive signal processing requirements [2].   While this field represents an area of current intensive research, some concepts, such as the systolic computing cellular array, have already produced significant developments [3].

The systolic array and related wavefront processor are specifically designed to exploit the unique regularities of a particular linear algebra operation [4].   In particular, a characteristic of many matrix algebra operations is the requirement for data communication between only nearest neighbour arithmetic cells in a properly designed array of such cells.   This basic principle of simplification in conjunction with the relatively low cost of VLSI arithmetic cells components makes it feasible to design hardwired systolic array algorithms with essentially no internal control, minimal memory, and maximal parallelism.   As a case in point, this paper discusses the broadband minimum variance distortionless response (MVDR) beamforming algorithm that consists of the following three matrix operations: Cholesky factorization of an estimated cross–spectral density matrix (CSDM), solution of a least–squares (filter) problem after each rank one update of the CSDM, and an N–channel matrix filter operation.

First, the theory and direct element level implementation of the adaptive broadband MVDR beamformer is presented; then the systolic array implementation is described. This is followed by a discussion of the theoretical performance predictions for an MVDR process.   Finally, some implications of systolic array architectures with respect to variations of the MVDR algorithm for high resolution space–time processing in very large arrays are considered.

6.2 THEORY AND DIRECT IMPLEMENTATION

A convenient discrete frequency–domain representation of the broadband data from an N–sensor array is given by the vector $\underline{x}_k$ with transpose $\underline{x}_k^T$ specified by

$$\underline{x}_k^T = [x_{1k}(\omega_1) \cdots x_{Nk}(\omega_1)x_{1k}(\omega_2) \cdots x_{Nk}(\omega_2) \cdots x_{1k}(\omega_M) \cdots x_{Nk}(\omega_M)]$$

$$= [\, \underline{x}_{1k}^T \, \underline{x}_{2k}^T \cdots \underline{x}_{Mk}^T \,] \quad , \tag{1}$$

where

$$x_{ik}(\omega_m) = \frac{1}{\sqrt{T}} \int_{t_k - \frac{T}{2}}^{t_k + \frac{T}{2}} x_i(t) \, e^{-j\omega_m t} \, dt \quad , \tag{2}$$

with $x_i(t)$ the output time–domain waveform of the i–th sensor and ω_m the m–th radial frequency

$$\omega_m = 2\pi m/T \quad 0 \leq m \leq M-1 \quad . \tag{3}$$

Therefore, the elements of the vector $\underline{x}_k$ are seen to be the discrete Fourier coefficients of $x_i(t)$ over the interval $(t_k - T/2,\ t_k + T/2)$. When the observation time, T, is large with respect to the inverse bandwidth of $x_i(t)$, the Fourier coefficients between frequencies become uncorrelated. We shall assume that this condition is satisfied and that the waveforms $x_i(t)$, $1 \leq i \leq N$, while not necessarily Gaussian random processes, are zero–mean and wide–sense stationary.

Consistent with the above waveform assumptions, the covariance matrix $R_m = R(\omega_m)$ at frequency ω_m, hereafter referred to as CSDM, is defined with ij–th element $E[x_{ik}(\omega_m)\ x_{jk}^*(\omega_m)]$. The covariance matrix for the Fourier coefficient vector $\underline{x}_k$ is an MN–by–MN block diagonal matrix with the CSDM R_m as the m–th N–by–N main block diagonal element. The MVDR beamformer filter $\underline{w}_k$ at time t_k is now defined in terms of the complex conjugate transpose (indicated by superscript H) as

$$\underline{w}_k^H = [w_{1k}^*(\omega_1) \cdots w_{Nk}^*(\omega_1)w_{1k}^*(\omega_2) \cdots w_{Nk}^*(\omega_2) \cdots w_{1k}^*(\omega_M) \cdots w_{Nk}^*(\omega_M)]$$

$$= [\underline{w}_{k1}^H \ \underline{w}_{k2}^H \cdots \underline{w}_{kM}^H] \quad . \tag{4}$$

The total broadband output power of the MVDR beamformer is defined as the variance

$$\sigma^2 \;=\; E[\,|\,\underline{w}_k^H\,\underline{x}_k|^2\,] \tag{5}$$

$$=\; \underline{w}_k^H\,E[\underline{x}_k\,\underline{x}_k^H]\,\underline{w}_k$$

$$=\; \sum_{m=1}^{M} \underline{w}_{mk}^H\,R_m\,\underline{w}_{mk} \;\;.$$

The requirements for a distortionless spectral response to a signal with the steering vector $\underline{d}_{mp}$ at frequency ω_m is equivalent to the constraint

$$\mathrm{Re}[\underline{w}_{mk}^H\,\underline{d}_{mp}] = 1 \quad \text{and} \quad \mathrm{Im}[\underline{w}_{mk}^H\,\underline{d}_{mp}] = 0, \quad 1 \le m \le M \tag{6}$$

on the MVDR filter weight vector $\underline{w}_{mk}$. The steering vector $\underline{d}_{mp}$ has the n–th element $\exp(j\omega_m\tau_{np})$ where τ_{np} is the relative time delay to the n–th sensor for a signal from the p–th direction. Using the method of undetermined Lagrange multipliers, σ^2 of Eq. (5) is minimized subject to the constraints of Eq. (6) if the criterion function

$$v \;=\; \sum_{m=1}^{M} [\underline{w}_{mk}^H R_m \underline{w}_{mk} \;+\; 2\lambda_{Rm}(\mathrm{Re}[\underline{w}_{mk}^H\underline{d}_{mp}] -1) \;+\; 2\lambda_{Im}(\mathrm{Im}[\underline{w}_{mk}^H\underline{d}_{mk}])] \tag{7}$$

is minimized with respect to the real and imaginary parts of $\underline{w}_{mk}$, simultaneously. The solution to this problem is known to be

$$\underline{w}_{mpk} \;=\; R_m^{-1}\,\underline{d}_{mp}^H / \underline{d}_{mp}^H\,R_m^{-1}\,\underline{d}_{mp} \;, \quad 1 \le m \le M \;, \tag{8}$$

where the beam index p indicates that a unique beam/filter vector $\underline{w}_{mpk}$ is required for each of the B $(1 \le p \le B)$ beam steering signal directions. The corresponding minimum total beamformer broadband output power (variance) obtained using Eq. (8) in Eq. (5) is

$$b_p^2 \;=\; \sum_{m=1}^{M} (\underline{d}_{mp}^H\,R_m^{-1}\,\underline{d}_{mp})^{-1} \;\;. \tag{9}$$

Therefore, the estimated signal autopower spectral density at frequency ω_m is

$$b_{mp}^2 = (\underline{d}_{mp}^H \ R_m^{-1} \ \underline{d}_{mp})^{-1} \ . \tag{10}$$

In practice, the CSDM matrix cannot be estimated exactly by time averaging because the random process $x_n(t)$ is never truly stationary and/or ergodic. As a result, the available averaging time is limited. Accordingly, one approach to the time–varying adaptive estimation of R_m at time t_k is to compute the exponentially time–averaged estimator of the CSDM R_m at time t_k as

$$R_{mk} = \mu R_{m,k-1} + (1 - \mu) \ \underline{x}_m \ \underline{x}_m^H \ , \tag{11}$$

where μ is a smoothing factor $(0 < \mu < 1)$ that implements the exponentially weighted time–averaging operation. Eq. (11) is a rank one update time–averaging operation to the CSDM at frequency ω_m and requires approximately $MN^2/2$ complex multiply addition operations and an equivalent memory size. The estimator R_{mk} is used in Eqs. (8), (9) and (10) to compute the MVDR filter weights, broadband beamformer output power, and estimated signal spectrum, respectively. A straightforward inversion of R_{mk} requires on the order of MN^3 operations. Approximately BMN^2 additional operations are required to form the beam filter vectors in Eq. (8) and beampower estimates in Eq. (9) for a system with B beams. The average number of operations for the direct implementation of the broadband MVDR beamformer is, therefore, on the order of

$$O_d = M((N^2/2) + BN^2 + (BN^2 + N^3)/K)$$

$$\simeq MN^2 (B + (B + N)/K) \tag{12}$$

per beamformer update cycle. The parameter K is the number of update cycles of the estimated CSDM R_m between computations of the updated filter weights and beam output power estimates.

6.3   SYSTOLIC IMPLEMENTATION

In terms of systolic computing array functions described in the following, implementation of the MVDR beamformer reduces the matrix inversion operations from order MN^3 to order MN^2. Systolic computing exposes the regularity of a particular function in such a way as to make that function amenable to the maximum of parallel computation and a minimum of both control and data transfer. In the following, the primary systolic computing elements are developed.

First, consider the requirement to compute the estimated output power for the m–th frequency, p–th beam steering direction, and k–th update cycle from the beam output power estimate

$$b_{mpk} = 1/g_{mpk}$$

where

$$g_{mpk} = \underline{d}_{mp}^H \, R_{mk}^{-1} \, \underline{d}_{mp} \; . \tag{13}$$

In the systolic implementation to be described, the inverse CSDM R_{mk}^{-1} is never formed explicitly. Rather the upper triangular Cholesky factor U_{mk} of R_{mk}, defined by

$$R_{mk} = U_{mk}^H \, U_{mk} \; , \tag{14}$$

is obtained directly (as explained later). The diagonal elements of the Cholesky factor are real. Substituting Eq. (14) into (13) yields

$$g_{mk} = ((U_{mk}^H)^{-1} \, \underline{d}_{mp})^H \, ((U_{mk}^H)^{-1} \, \underline{d}_{mp})$$

$$= \underline{v}_{mpk}^H \, \underline{v}_{mpk}$$

$$= |\underline{v}_{mpk}|^2 \; , \tag{15}$$

where only the complex vector $\underline{v}_{mpk} = (U_{mk}^H)^{-1} \, \underline{d}_{mp}$ must be computed at a cost of $MN^2/2$ operations. To form the beam output from the beam filter vector of Eq. (8), i.e.,

$$y_{mpk} = \underline{w}_{mpk}^H \, \underline{x}_{mk}$$

$$= \underline{d}_{mp}^H \, R_{mk}^{-1} \, \underline{x}_{mk}/g_{mpk} \; , \tag{16}$$

again using the Cholesky factorization of Eq. (14) to obtain

$$y_{mpk} = ((U_{mk}^H)^{-1} \, \underline{d}_{mp})^H \, ((U_{mk}^H)^{-1} \, \underline{x}_{mk})/g_{mpk}$$

$$= \underline{v}_{mpk}^H \, \underline{z}_{mpk} \, / \, |\underline{v}_{mpk}^H|^2 \; . \tag{16}$$

The important point with respect to Eq. (16) is that the computation of the vector

$$\underline{z}_{mpk} = (U_{mk}^H)^{-1} \underline{x}_{mk} \tag{17}$$

is of the exact form as that for

$$\underline{v}_{mpk} = (U_{mk}^H)^{-1} \underline{d}_{mp} \tag{18}$$

in Eq. (15).   Thus, the same circuitry can be used for each computation.

A realization of the MVDR beamformer as described above requires the implementation of two functions with systolic arrays: a rank one update of the CSDM R_{mk} in terms of its Cholesky factor U_{mk} in Eq. (14) and the subsequent solution of a set of N linear equations of the form

$$U_{mk}^H \underline{h} = \underline{b} \tag{19}$$

as required by Eqs. (17) and (18).

6.3.1   RANK ONE CHOLESKY FACTOR UPDATE [5,6]

Let the N–dimensional row vector

$$\underline{z} = (1 - \mu)^{1/2} \underline{x}_m^H \tag{20}$$

$$= [z_1 \ z_2 \ ... \ z_N] \tag{21}$$

and the upper triangular N–by–N matrix

$$U = \mu^{1/2} U_{m,k-1} \tag{22}$$

be defined.   Now form the (N + 1)–by–N data augmented Cholesky matrix

$$B = \left[\begin{array}{c} \underline{z} \\ \hline U \end{array} \right]$$

$$= \left[\begin{array}{cccc} z_1 & z_2 & \cdots & z_N \\ u_{11} & u_{12} & \cdots & u_{1N} \\ 0 & u_{22} & \cdots & u_{2N} \\ \vdots & \vdots & & \\ 0 & 0 & \cdots & u_{NN} \end{array} \right] , \tag{23}$$

which can be reduced to upper triangular form by a sequence of N orthogonal plane rotations indicated by the orthogonal premultiplier matrix P as

$$PB = \left[\begin{array}{c} \bar{U} \\ \hline \underline{0}^T \end{array} \right] \qquad (24a)$$

$$= \left[\begin{array}{cccc} \bar{u}_{11} & \bar{u}_{12} & \cdots & \bar{u}_{1N} \\ 0 & \bar{u}_{22} & \cdots & \bar{u}_{2N} \\ \vdots & \vdots & & \vdots \\ 0 & 0 & \cdots & \bar{u}_{NN} \\ 0 & 0 & \cdots & 0 \end{array} \right], \qquad (24b)$$

where $P = P_N \, P_{N-1} \cdots P_1$, $P_n^H \, P_n = I$ and $\underline{0}$ is an N–vector of all zeroes. Thus,

$$B^H P^H PB = B^H B \qquad (25a)$$

$$= U^H U + z^H z \qquad (25b)$$

$$= \mu U_{m,k-1}^H \, U_{m,k-1} + (1 - \mu) x_m \, x_m^H \qquad (25c)$$

$$= R_{mk} \qquad (25d)$$

$$= U_{mk}^H \, U_{mk} \qquad (25e)$$

$$= \bar{U}^H \bar{U} \qquad (25f)$$

and U is the rank one updated Cholesky factor of the estimated CSDM R_{mk}. If u_{ii}, $1 \leq i \leq N$ is real, then u_{ii}, $1 \leq i \leq N$ is also real when the boundary cell process definition of figure 1 is used.

The systolic array realization of the rank one Cholesky factor update requires both a boundary cell and an internal cell as illustrated in figure 1. The specific linear systolic array for reduction of Eq. (23) to upper triangular form is given in figure 2.   It is the boundary cell that must compute the correct plane rotation angle complex cosine and sine values (c,s) and the internal cells which propagate that rotation down the n–th row of the matrix $P_{n-1} \, P_{n-2} \cdots P_1 \, U$.

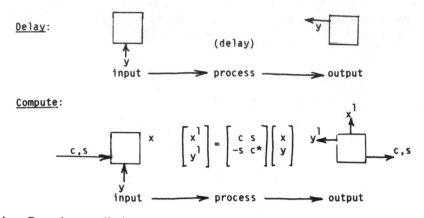

(a)   Boundary cell input–process–output diagram for Cholesky Factor update (real y).

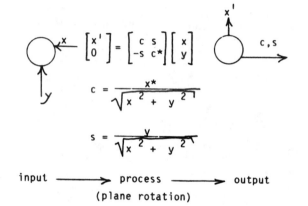

(b)   Internal cell input–process–output diagram for the delay and compute function required for Cholesky Factor update.

Figure 1   Boundary cell and internal cell input–process–output diagram for the Cholesky Factor update.

6.3.2 LINEAR EQUATION BACK SOLVE [7]:

The solution of Eq. (19) is performed as a sequential backsubstitution of the elements of h as they are obtained in sequence.   Eq. (19) is of the form

$$
\begin{bmatrix}
u_{11} & 0 & 0 & \cdots & 0 \\
u_{12} & u_{22} & 0 & \cdots & 0 \\
u_{13} & u_{23} & u_{33} & \cdots & \cdot \\
\vdots & \vdots & \vdots & \ddots & \vdots \\
u_{1N} & u_{2N} & u_{3N} & \cdots & u_{NN}
\end{bmatrix}
\begin{bmatrix}
h_1 \\
h_2 \\
h_3 \\
\vdots \\
h_N
\end{bmatrix}
=
\begin{bmatrix}
b_1 \\
b_2 \\
b_3 \\
\vdots \\
b_N
\end{bmatrix}
\tag{26}
$$

$\bar{u}_{11}$

. $\bar{u}_{12}$

$\bar{u}_{22}$. $\bar{u}_{13}$

. $\bar{u}_{23}$. $\bar{u}_{14}$

$\bar{u}_{33}$. $\bar{u}_{24}$.

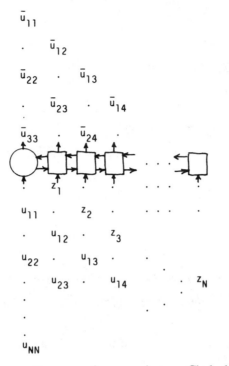

z_1

u_{11} . z_2 .   .   .   .   .

. u_{12} . z_3

u_{22} . u_{13} .   .

. u_{23} . u_{14} . z_N

.

.

u_{NN}

Figure 2   Linear systolic array for a rank–one Cholesky Factor update.

Thus

$$h_1 \;=\; b_1/u_{11} \tag{27}$$

$$h_2 \;=\; (b_2 - u_{12}h_1)/u_{22}$$

$$h_3 \;=\; (b_3 - u_{13}h_1 - u_{23}h_2)/u_{33}$$

$$\vdots$$

$$h_N \;=\; (b_n - u_{1N}h_1 - \cdots - u_{N-1,N}h_{N-1})/u_{NN} \;\;,$$

which is realized with an implementation of the linear systolic array using the cells illustrated in figure 3. The corresponding systolic array is shown in figure 4.

It is noted that systolic arrays for the rank one Cholesky update (figure 2) and linear equation backsolve solution (figure 4) are linear arrays with minor differences in input–output structure. As discussed above, utilization of the boundary and internal cells is fifty percent. One hundred percent utilization is achieved with the interleaving of two problems delayed by one cycle. Moreover, it is possible using a nonalgebraic, i.e., geometric approach, to partition large (N) problems into segments wherein a smaller systolic array of size less than N internal cells can be time multiplexed.

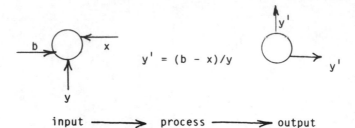

(a)   Boundary cell input–process–output diagram for linear equation backsubstitution.

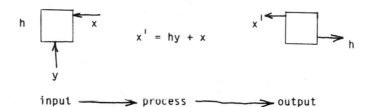

(b)   Internal cell input–process–output diagram for linear equation backsubstitution.

Figure 3   Linear equation backsubstitution.

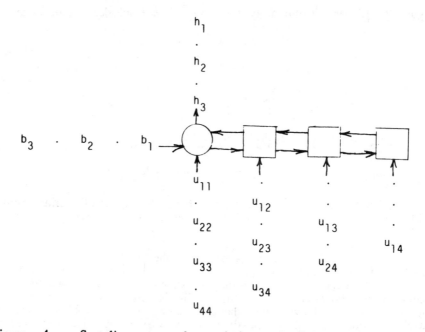

Figure 4   Systolic array for solution of linear equations by backsubstitution for a 4–element partition.

6.3   PERFORMANCE OF AN MVDR BEAMFORMER

The CSDM for an array that receives both a signal with autopower spectral density (ASD)σ_s^2 and an interference, which is uncorrelated with the signal, and an ASD σ_i^2 at frequency f is given by

$$R = \sigma_s^2 \, \underline{d}_s \, \underline{d}_s^H + \sigma_i^2 \, \underline{d}_i \, \underline{d}_i^H + \sigma_o^2 \, I \ . \tag{28}$$

In Eq. (28), σ_o^2 is the spatially uncorrelated noise ASD at frequency ω, I is an N–by–N identity matrix, and $\underline{d}_s$ and $\underline{d}_i$ are the steering vectors for the signal and interference, respectively.   The interference is assumed to be in the farfield of the array, i.e., it is a Point INterference (PIN) with respect to angle of arrival.

The MVDR filter vector for beamforming in the signal direction at frequency ω has been shown to be

$$\underline{w} = R^{-1} \, \underline{d}_s / \underline{d}_s^H \, R^{-1} \, \underline{d}_s \ . \tag{29}$$

For a conventional time delay–and–sum beamformer (CBF) the vector $\underline{d}_s$ is used.   The gain in signal to the total of interference and noise power at the beam output relative to a single sensor is given by

$$G_{MV} = (\sigma_s^2 |\underline{w}^H \underline{d}_s|^2 / \underline{w}^H [\sigma_i^2 \underline{d}_i \underline{d}_i^H + \sigma_o^2 \tau] \underline{w}) / (\sigma_s^2 / [\sigma_i^2 + \sigma_o^2]) \tag{30}$$

for the MVDR beamformer and

$$G_c = (\sigma_s^2 N^2 / \underline{d}_s^H [\sigma_i^2 \underline{d}_i \underline{d}_i^H + \sigma_o^2 \tau] \underline{d}_s) / (\sigma_s^2 / [\sigma_i^2 + \sigma_o^2]) \tag{31}$$

for the CBF.   The array gain improvement (AGI) of the MVDR relative to the CBF is

$$AGI = G_{MV}/G_c$$

$$= 1 + \frac{r^2 \, \ell(1-\ell)}{1+r} \tag{32}$$

where

$$r = N\sigma_i^2 / \sigma_o^2 \tag{33}$$

and represents the interference to uncorrelated noise variance ratio at the ouput of a CBF steered with its main response axis (MRA) directly at the interference. The parameter ℓ ($0 \leq \ell \leq 1$) is defined by the expression

$$\ell = |\underline{d}_s^H \underline{d}_i|^2 / N^2 \quad . \tag{34}$$

Figure 5 gives the AGI as a function of r for four different values of ℓ; namely, $\ell = 1/2$ (-3 dB), $\ell = 2/100$ (-17 dB), $\ell = 5/1000$ (-23 dB) and $\ell = 1/1000$ (-30 dB). For either high interference levels and/or low noise levels, significant gains of greater than 10 dB are achievable with MVDR beamforming over a CBF. At high noise levels the increased uncorrelated noise begins to mask the PIN as the major degrading factor and useful AGI is only achieved when the PIN is inside the mainlobe of the beam receiving the signal (-3 dB re MRA).

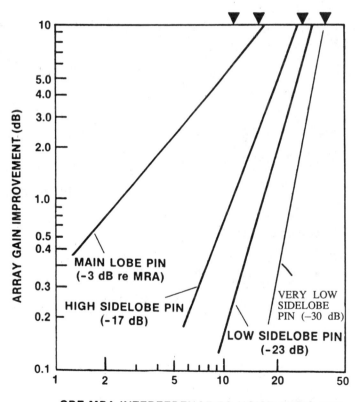

Figure 5 Minimum variance optimum beamformer AGI relative to a CBF as a function of the interference–to–noise ratio for a PIN measured at the output of a CBF steered directly at the interference.

6.5   ADAPTIVE BEAMFORMING FOR VERY LARGE ARRAYS

The fundamental parameter which determines the signal to total background noise variance ratio gain for an array is the number of sensors N. Simply stated, if an array cannot provide a certain level of required conventional beamformer array gain with only spatially uncorrelated white noise present, then adaptive beamforming will not alter this fact and the only option is to make the number of sensors (N) in the array larger. To illustrate this point, it is observed that

$$G_{MV}\big|_{\sigma_i^2=0} \;=\; G_C\big|_{\sigma_i^2=0} \;=\; N \;. \tag{35}$$

Thus, from the array gain perspective, CBF is optimum when there is no interference present and the only recourse is to build larger arrays, ie., arrays with more sensors.

Given the specific spatially uncorrelated white noise array gain, N, of Eq. (35) for an array of N sensor elements, an implementation of the MVDR beamformer described previously requires the update of an N dimensional Cholseky factor and the solution of a correspondingly large set of linear equations. Thus, the number of sensor elements N equal to the array white noise gain determines the size of the element space MVDR system. For arrays with a large number of sensor elements N, the computational requirements can become prohibitive given that the computing burden is proportional N^3 as specified by Eq. (12).

In a dynamic situation, where the angle of arrival for a particular interference is changing with time, the effective averaging time for estimating the interelement CSDM Cholesky factor is limited by a temporal stationarity assumption. Thus, the variance of the elements in the CSDM estimator of Eq. (11), which is inversely proportional to averaging time, has a lower bound determined by the finite averaging time. Specifically, if M is the effective number of statistically independent sample vectors, $\underline{x}_m$, which are exponentially averaged to produce the estimated CSDM of Eq. (11), then the variance on the MVDR beam output power estimator detection statistic of Eq. (10) is inversely proportional to M−N+1 where it is assumed that M > N   [8,9]. Thus, as the number of elements (N) in the array increases, it is necessary to increase the CSDM estimator averaging time as determined by M proportionately to maintain the same beam output power estimator variance.

Eventually, for the element space MVDR process, the size of the array N is limited by the time stationarity constraint which is, in turn, determined by the interference position rate of change with respect to time.

A natural way to avoid the temporal stationarity limitation on the white noise array gain N discussed above is to perform the systolic MVDR process in a domain other than the N–dimensional element space. If this new domain has a lower dimensionality, both the systolic engine computational and memory size complexity and the effective time averaging requirement constraints are reduced accordingly. References [10] and [11] suggest the implementation of the MVDR algorithm in a so–called beam space. In beam space, only spatially orthogonal CBF beams that are steered contiguous to a selected reference beam are used as inputs to an MVDR beam interpolation algorithm. Clearly, the question of selecting the appropriate number of and location for orthogonal beams is not straightforward. At a minimum, this selection is a frequency dependent process due to the variation of beam overlap caused by the increase of beamwidth with a decrease in frequency. In addition, the number of independent beams must be made large enough to provide a sufficient number of degrees of freedom for near optimum performance in a multiple interference condition.

As a practical matter, even the formation of a conventional time delay–and–sum beam for a very large array is a difficult implementation issue. Usually partial aperture, ie., subarray, beams are formed as a first step in the formation of a full beam from a large array. An alternative to the beam space approach for dimensionality reduction in very large arrays is referred to herein as the subarray (SA) space formulation [12,13]. In this approach, subapertures of contiguous elements in the large array are prebeamformed using simple time delay–and–sum and fixed spatial windowing techniques. This partitioned subarray beamforming (SBF) can be envisioned as creating a secondary array of spatially directional elements which, in turn, are processed with an N/P–dimension MVDR beamformer in cascade with the subaperture beamformers discussed above.

In each of the partitioned subarrays (N/P), as illustrated in Figure 6, consists of P contiguous sensor elements, then the MVDR process is of dimension N/P. However, as with the beam space approach, more than one small (dimension N/P) CSDM Cholesky factor needs to be estimated at each frequency. This is in contrast with the element space MVDR where a single very large (dimension N) CSDM Cholesky factor is estimated. This is because each SBF can form approximately P spatially independent beams which resolve substantially nonoverlapping segments of solid angle. Thus, for the formation of a particular CSDM matrix estimator, those SBF outputs steered at the same angle should be selected. This SA space requirement would constitute a need for approximately P MVDR parallel processes, each of dimension N/P as opposed to one MVDR process of dimension N required for the element space formulation. It is noted that from the computational requirement standpoint, the SA space burden is proportional to $B_{SA} = (N/P)^2 N$ as contrasted to $B_E = N^3$ for element space. The actual burden is proportional to $[(N/P)^2 P + (N/P)^2 N] = (N/P)^2 N$ for $N \gg P$. The Cholesky factor update burden is $(N/P)^2 P$ and the backsubstitution burden is $(N/P)^2 N$. Thus, the computational load and memory size reductions can be enormous when the SA space is adopted. Moreover, the restriction on the effective averaging time M imposed by the array size N becomes $M-(N/P)+1 \geq T$, where T is a threshold set by the desired variance of the beam output power detection statistic. It follows that averaging time can be reduced in a SA space formulation to accommodate the spatial dynamics of the interference with essentially no loss of performance.

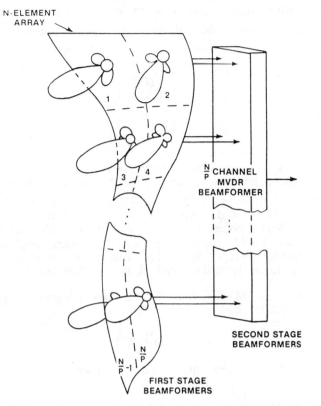

Figure 6 Subarray space partitioning and cascaded beamforming of a very large N–element array to reduce N/P–Input MVDR Beamformer complexity and convergence time restrictions. Each of (N/P) subarrays consists of P elements with conventional beamforming.

The beam and SA space MVDR array gain performance is obtained by introducing the array element data preprocessing matrix

$$
D = \begin{cases}
[\; d_{-L/2} \cdots d_0 \cdots d_{L/2} \;] \; \text{for beamspace} & (36a) \\[2mm]
\begin{bmatrix}
\underline{d}_{s1} & 0 & \cdots & 0 \\
0 & \underline{d}_{s2} & \cdots & 0 \\
\vdots & \vdots & \ddots & \vdots \\
0 & 0 & \cdots & \underline{d}_{s,N/P}
\end{bmatrix} \; \text{for SA space} \; . & (36b)
\end{cases}
$$

In Eq. (36a), $\underline{d}_k$ is a CBF N–dimensional steering vector corresponding to a beam space patch of size L+1 which is defined as being centered at a point, θ, specified by a reference beam which has steering vector $\underline{d}_0 = \underline{d}_0(\theta)$.

Ideally, these steering vectors are assumed to be orthogonal, i.e., $\underline{d}_i^H \underline{d}_j = N\delta_{ij}$. For the SA space MVDR process, the P–dimensional vector $\underline{d}_{sk}$ corresponds to the P–element subvector of the N–dimensional steering vector:

$$\underline{d}_o = \begin{bmatrix} \underline{d}_{s1} \\ \underline{d}_{s2} \\ \vdots \\ \underline{d}_{s,N/P} \end{bmatrix} , \qquad (37)$$

which would be required to electronically steer the entire array subarray by subarray at a single point that is implicit in the steering vector $\underline{d}_o$. For the beam space formulation D is an N–by–(L+1) dimensional matrix and for the subarray counterpart D is of dimension N–by–(N/P). If (L+1) = N/P for these two suboptimum MVDR processes and the same AGI performance results, then the two approaches would require the same hardware implementation with identical performance except that the beamspace process requires full conventional beams to be formed instead of only partial beams.

To establish the MVDR AGI for the two suboptimum procedures, the reduced dimension CSDM matrix

$$\bar{R} = D^H R D \qquad (38)$$

is defined. For both the beam space and SA space processes, the secondary beam output variance

$$\sigma_b^2 = \underline{\bar{w}}^H \bar{R} \underline{\bar{w}} \qquad (39)$$

is minimized with respect to either the (L+1) or (N/P) dimensional vector $\underline{w}$. The constraints that the element in location (L/2)+1 of $\underline{w}$ be unity for the beamspace and

$$N = \underline{\bar{w}}^H \underline{D}^H \underline{d} \qquad (40)$$

for the SA space procedures are required to satisfy the distortionless signal constraint.

It is a direct procedure to obtain the following AGI expression

$$AGI_b = 1 + \frac{r^2 \ell([L+1]\ell-\ell)}{1 + r[L+1]\ell} , \qquad (41a)$$

and

$$AGI_s = 1 + \frac{r^2 \ell(\ell_s-\ell)}{1 + r\ell s} \qquad (41b)$$

for the beam space and SA counterparts, respectively, to Eq. (32) which corresponds to the fully optimum element space configuration. In Eq. (41a),

$$[L+1]\ell \;=\; \sum_{k=-\frac{L}{2}}^{\frac{L}{2}} \ell_k \tag{42}$$

where $\ell_k = |\underline{d}_i^H \underline{d}_k|^2/N^2$ $(0 \le \ell_k \le 1)$ is the relative response level of the interference in the k–th beam output of the beam space patch. The quantity ℓ can be thought of as the interference response level averaged over all (L+1) beams in the beam space patch. In Eq. (41b), ℓ_s $(0 \le \ell_s \le 1)$ is the relative response level of the interference in the SA beam output. Note that in the limiting case for SA MVDR processing, P=1 corresponds to only one sensor per SA. Here the SA has an omnidirectional response so that $\ell_s = 1$ and the result is the same as Eq. (32) as expected. It is observed that for the two suboptimum MVDR processes to perform equally, then

$$e = \ell_s \tag{43}$$
$$= [L+1]\ell$$

and optimality is approached only to the extent that e approaches unity. Figure 7 gives the AGI metric

$$AGI(e) \;=\; 1 + \frac{r^2 \, \ell(e - \ell)}{1 + re}$$

for the suboptimum, reduced dimension beam and subarray space MVDR processes for several values of e. The same values of the interference response level, ℓ, for the full aperture CBF are used as in Figure 5. It is significant that at high interference–to–noise levels (r) the suboptimum procedures are nearly equivalent to the optimum element based process except for $\ell = 1/2$. Furthermore, it is primarily only for large r that substantial sidelobe interference AGI is obtained. For mainlobe PIN, when $\ell = 1/2$ the beamspace MVDR would have a substantial performance loss. This is because e only differs from 1/2 by the average sidelobe level of the interference over the remaining beams in the patch and this would be a small number. Thus, for interference within the mainlobe subarray MVDR would be superior.

6.6   SUMMARY AND CONCLUSIONS

The fundamentals of adaptive beamformer (ABF) implementation using systolic computing arrays have been presented. It has been shown that for a continuously updating ABF, a direct open loop realization can be obtained with

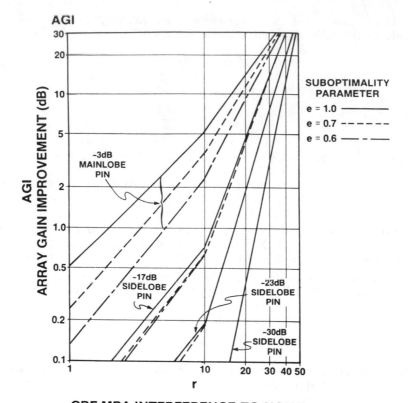

Figure 7 Comparison of (suboptimum subarray and beam space) and optimum (element space) MVDR process array gain improvement versus interference–to–noise ratio.

a linear systolic array consisting of just two types of functional computing cells. The performance of a generic ABF system has been reviewed. Finally, the problem of computational burden, memory requirements, and extreme convergence time associated with arrays having large number of elements has been addressed by showing that systolic array techniques need be applied only at the second stage of a cascaded beamformer. The tremendous saving in MVDR implementation hardware with application of the suboptimum processes could offset the loss of performance. The preferred suboptimum MVDR processes use subarray prebeamforming because it is extremely regular in its architecture; it is not frequency–dependent; and it yields better AGI performance for the same MVDR complexity.

6.7   REFERENCES

1.   S. Haykin, J. Justice, N. Owsley, and A. Kak, <u>Array Signal Processing</u>, Prentice–Hall,   1985.

2.   T. Kailath, S. Y. Kung, and H. Whitehouse, <u>VLSI and Modern Signal Processing</u>, Prentice–Hall,   1985.

3.   H.T. Kung, "Why Systolic Arrays", <u>IEEE Trans. Computers</u>, 15 (1), pp. 37–46, 1982.

4.   Y.S. Kung et al., "Wavefront Array Processor: Language, Architecture, and Applications," <u>IEEE Trans. Computers</u>, C–31, pp. 1054–1066, 1982.

5.   P. Kuekes, J. Avila, and D. Kandle, <u>Adaptive Beamforming Design Specification</u>, ESL Inc., Sunnyvale, CA, 20 May 1986.

6.   R. Schreiber and Wei–Pai Tang, "On Systolic Arrays for Updating the Cholesky Factorization", Royal Institute of Technology, Stockholm,Sweden, Dept. of Numerical Analysis and Comp. Science, TRITA–NA–8313, 1984.

7.   G. Strang, <u>Linear Algebra and It's Applications</u>,   Academic Press, 1980, Second Edition.

8.   J. Capon and N. Goodman, "Probability Distributions for Estimators of Frequency Wavenumber Spectrum", <u>Proc. IEEE</u>, Vol. 58, pp. 1785–1786, 1970.

9.   J. Capon, "Correction to Ref. [8],," <u>Proc. IEEE</u>, Vol. 59, p. 112, 1971.

10.   A.H. Vural, "A Comparative Performance Study of Adaptive Array Processors", <u>Proceedings of IEEE ICASSP</u>, Hartford, CT, 1977, pp. 695–700.

11.   D.A. Gray,   "Formulation of the Maximum Signal–to–Noise Ratio Array Processor in Beam Space," <u>J. Acoust. Soc. Am.</u>, 72 (4), October 1982, pp. 1195–1201.

12.   N.L. Owsley and J.F. Law,   "Dominant Mode Power Spectrum Estimation", <u>Proceedings of IEEE ICASSP</u>, Paris, April, 1982, Vol. 1, pp. 775–779.

13.   N.L. Owsley,   "Signal Subspace Based Minimum–Variance Spatial Array Processing", <u>Proceedings of Asilomar Conf. on Circuits, Systems, and Computers</u>, November 6–8, 1985, pp. 94–97.

7
CURRENT METHODS OF DIGITAL SPEECH PROCESSING

L.R. RABINER, B.S. ATAL, AND J.L. FLANAGAN
AT&T Bell Laboratories
Murray Hill, New Jersey

7.0 ABSTRACT

The field of digital speech processing includes the areas of speech coding, speech synthesis, and speech recognition. With the advent of faster computation and high speed VLSI circuits, speech processing algorithms are becoming more sophisticated, more robust, and more reliable. As a result, significant advances have been made in coding, synthesis, and recognition, but, in each area, there still remains great challenges in harnessing speech technology to human needs.

In the area of speech coding, current algorithms perform well at bit rates down to 16 kbits/sec. Current research is directed at further reducing the coding rate for high–quality speech into the data speed range, even as low as 2.4 kbits/sec. In text–to–speech synthesis we are able to produce speech which is very intelligible but is not yet completely natural. Current research aims at providing higher quality and intelligibility to the synthetic speech produced by these systems. Finally, in the area of speech and speaker recognition, present systems provide excellent performance on limited tasks; i.e., limited vocabulary, modest syntax, small talker populations, constrained inputs, and favourable signal–to–noise ratios. Current research is directed at solving the problem of continuous speech recognition for large vocabularies, and at verifying talker's identities from a limited amount of spoken text.

7.1 INTRODUCTION

Although the field of speech processing is quite broad and encompasses a number of diverse application areas, we will be concerned in this paper only with the areas of speech coding, synthesis, and recognition. For each of these important application areas we will review the present status of the technology and discuss the directions in which research is heading.

Speech coding is concerned with communication between people and therefore deals with techniques of speech transmission, generally over the conventional telephone network. Of central concern here are methods for reducing the required bandwidth (or equivalently the digital bit rate) for transmitting speech. Speech synthesis, or computer voice response as it is often called, is concerned with machines talking to people. Although systems as simple as announcement machines fall into this area, we will primarily be concerned with the state of the art and current research directions in the area of text–to–speech synthesis systems. Speech recognition is concerned with people talking to machines. Speech recognizers range in sophistication from the simplest isolated word/phrase recognition systems, to fully conversational recognizers that attempt to deal with vocabularies and syntax comparable to natural language. Also included in the broad area of speech recognition is the topic of speaker recognition in which the job of the machine is either to verify the claimed identity of a talker, or to identify the individual talker as one of a fixed, known population.

Digital speech processing has advanced in the past few years for several reasons. One key reason is the explosive growth in computational capabilities, supported by economical VLSI hardware. General–purpose signal processing computers exist today, which can run standard programming languages and can execute algorithms at rates on the order of 50–200 megaflops [1]. Such machines are classified as mini–supercomputers, and cost less than main–frame machines of a few years past. Similarly, VLSI digital speech processor (DSP) chips now exist which do calculations in floating–point arithmetic at an 8 megaflop rate [2]. Thus even a 100 megaflop algorithm can potentially be realized with about a dozen DSP chips on a single circuit board.

Another reason for the progress in digital speech processing is the improvements that have been made in speech processing algorithms. Speech coding has benefited significantly from the introduction of MPLPC (multi–pulselinear predictive coding) [3], and CELP (code–excited linear prediction) [4]; the field of text–to–speech synthesis has seen major improvements due to the introduction of large pronouncing dictionaries; and the field of speech recognition has seen the maturity of algorithms for recognizing connected words (e.g., level building [5]), and the widespread acceptance of statistical modeling techniques (namely hidden Markov models or HMM's) [6,7].

Finally, perhaps the greatest recent impetus in advancing digital speech processing has been the growing need for products that serve real–world applications. The past decade has seen major growth in the utility of voice products for at least four market sectors – namely, telecommunications, business applications, consumer products, and government. In the telecommunications sector, voice coders are used for reduced bit–rate transmission and privacy; repertory name dialers are used for hands–free dialing; announcement systems are used to speak computer stored information to customers; and a wide variety of operator and attendant services depend upon recognition and synthesis for increased utility. In business applications, voice mail and store–and–forward services are already in widespread use, and voice interactive terminals and workstations are beginning to appear on the market. In the consumer products and services sector, toys using either synthesis and/or recognition have been available for several years, and recently residence communication systems and alarm announcement systems have started to appear. In the area of government communications, anticipated uses include coding for secure communications, and voice control of military systems.

The above examples, by no means exhaustive, illustrate the burgeoning applications of speech processing and point to a growing market in the coming years.

It is the purpose of this paper to outline the main issues in speech coding, synthesis, and recognition, to indicate where progress has ben made, and to point out areas where new research is necessary to achieve desired goals.

7.2   SPEECH CODING

In the new emerging digital communication environment, transmission of digital speech at low bit rates without compromising voice quality is becoming increasingly important.   Low bit rate voice will play a key role in providing new capabilities in future communication systems — e.g., for sending voice mail over telephone networks, for integrating voice and data in packet systems for transmission, for narrow—band cellular radio, and for insuring privacy in voice communication.

The speech coding technology to achieve high voice quality is well developed for bit rates as low as 16 kbits/sec [8].   The major research action is now focused at bringing the rate significantly lower than 16 kbits/sec without seriously degrading the speech quality.   The lower bit rates facilitate end—to—end digital voice communication over dialed—up pubic telephone lines, and are important to spectrum conservation in mobile radio.

Real—time implementation of low—bit—rate—voice coders previously has been a difficult and costly task.   Recent advances in device technology and the availability of fast programmable digital signal processors [9] has made the task easier.   We are now able to implement fairly complex speech processing algorithms on a single chip [10].

7.2.1   PRESENT SPEECH CODING TECHNOLOGY

The objective in speech coding is to transform the analog speech signal to a digital representation.   Redundancies, introduced in the speech signal during the human speech production process, make it possible to encode speech at low bit rates. Moreover, our hearing system is not equally sensitive to distortions at different frequencies and has a limited dynamic range.   Speech coding techniques take advantage of these properties for reducing the bit rate. We can summarize the present status of our capabilities for transmitting high quality speech at low bit rates.

Figure 1 shows the variation of speech quality versus transmission bit rate for three coder technologies.   Typically, performance of speech coders diminishes with decreasing transmission rate.   In Fig. 1 the speech quality is expressed on a scale which includes the terms excellent, good, fair and poor. Often speech quality is expressed in terms of a Mean Opinion Score (MOS) on a 5—point scale where an MOS of 5 is excellent quality, 4 is good, 3 is fair, 2 is poor, and 1 is unsatisfactory.

As shown in Fig. 1, two traditional coder technologies are waveform coders and vocoders. Waveform coders aim at reproducing the speech waveform as faithfully as possible.   They provide high—quality speech above 16 kbits/sec but their performance usually falls off rapidly at much lower bit rates. Vocoders use a model of human speech production to obtain a more efficient representation of the speech signal, and thus are able to bring the bit rate

BIT RATE VERSUS SPEECH QUALITY FOR SPEECH CODERS

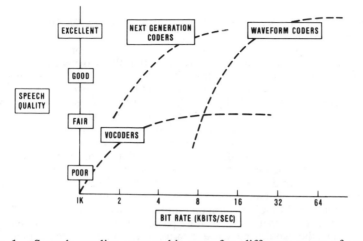

Figure 1   Speech quality versus bit rate for different types of coders.

down to much lower values – even as low as 400 bits/sec – but, with present understanding, the speech quality is significantly impared.   Our ability to provide high–quality speech below 16 kbits/sec is limited at present, but the next generation of coders, taking advantage of new capabilities offered by VLSI technology as well as new understanding in speech coding, promise to fill this gap in performance.

Speech coding methods have been standardized both at 64 and 32 kbits/sec and coders at these rates are being used in the public switched telephone networks.   There are no published civil standards yet for lower bit rates, although there exists a military standard for a 2.4 kbit/sec vocoder.

The bit rate of 16 kbit/sec is suitable for a variety of applications, such as voice mail, secure voice over wide–band cellular radio channels, and integrated transmission of voice and data over packet networks.   The coding technology to achieve high quality at 16 kbits/sec is available at present. These coding techniques are more complex in comparison to ones used in standard PCM and ADPCM coders, but they can be implemented on a single digital signal processor chip to perform real–time coding.   Adaptive predictive coders [11], sub–band coders with adaptive bit allocation [12], and multi–pulse linear predictive coders [13] are a few examples of coders capable of providing high–quality speech at 16 kbits/sec.   These coders have been implemented on a single DSP chip, and subjective tests based on these implementations provide a mean opinion score (MOS) of 3.9 for the multi–pulse coder, and a 3.8 for the adaptive bit allocation sub–band coder, both operating at 16 kbits/sec.   For comparison, standard mu–law PCM (56kbps) and ADPCM (32 kbps) coders have MOS of 4.5 and 4.0, respectively.   Another hybrid coder that combines the adaptive predictive and multi–pulse coders has produced speech at 16 kbits/sec with quality exceeding that of the ADPCM coder at 32 kbits/sec [14].   A multi–pulse coder is capable of providing high–quality speech at rates even lower than 16 kbits/sec, and details of this technique are discussed next.

7.2.2 SPEECH SYNTHESIS MODELS FOR LOW BIT RATE CODING

A proper speech synthesis model, capable of reproducing many different voices and requiring a small amount of control information, is essential for achieving high voice quality at low bit rates. A synthesis model that has been popular over many years is the traditional vocoder model where the synthetic speech is generated by exciting a linear filter with pitch pulses or white noise. The limitations of this simple vocoder model are now well known. The multi–pulse LPC model [15] seeks to overcome such limitations by replacing the traditional pitch pulse and white noise excitation with a sequence of pulses whose amplitudes and locations are chosen to minimize the perceptual difference between original and synthetic speech signals. Figure 2 illustrates both the traditional vocoder and the multi–pulse excitation models. The multi–pulse model has enough flexibility to reproduce a wide variety of speech waveforms, including voiced and unvoiced speech. The model is reasonably efficient in that only a few pulses (typically 8 to 16 every 10 msec) are needed in the multi–pulse excitation to produce high–quality synthetic speech. Further reduction in the number of pulses, in particular for high–pitched voices, can be achieved by incorporating a linear filter with a pitch loop in the synthesizer [13].

Recently, another model, based on stochastic excitation [15], has shown great promise for producing high–quality speech at low bit rates. In this model, the excitation is selected from a codebook of random white Gaussian sequences using a fidelity criterion that minimizes the perceptual difference between the original and synthetic speech signals. The different synthesis models are illustrated in Fig. 3. Both multi–pulse and stochastic models use identical linear filters to introduce correlations at short and long delays in the output speech signal, but they differ mainly in the manner in which the excitation to the linear filter is specified.

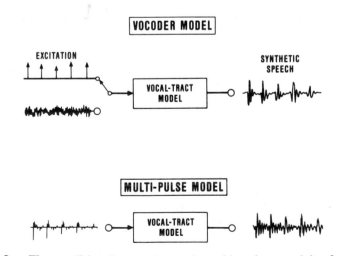

Figure 2   The traditional vocoder and multi–pulse models for speech synthesis.

SPEECH SYNTHESIS MODELS

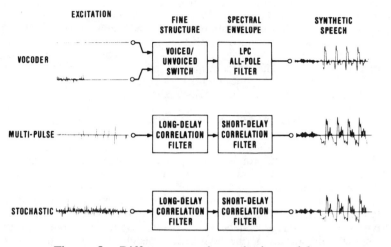

Figure 3   Different speech synthesis models.

7.2.3   EXCITATION MODELS FOR LOW BIT RATE VOICE

The principle for determining the excitation in multi–pulse coders is illustrated in Fig. 4.   The synthetic speech signal at the output of the synthesis filter is compared with the original speech signal and the error signal is further processed to produce a measure of perceptual error.   This processing includes linear filtering of the objective error to attenuate those frequencies where the error is perceptually less important and amplify those frequencies where the error is perceptually more important.   The excitation is chosen to minimize the perceptual error.

MULTI-PULSE EXCITATION ANALYSIS PROCEDURE

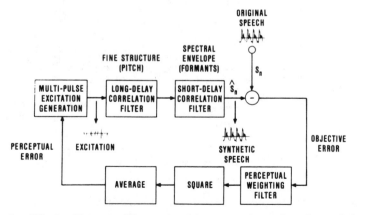

Figure 4   Block diagram illustrating the procedure for determining the optimum excitation in multi–pulse and stochastic coders.

The locations and amplitudes of the pulses in the multi–pulse excitation are obtained sequentially — one pulse at a time. After the first pulse has been determined, a new error is computed by subtracting out the contribution of the first pulse to the error and the location of the next pulse is determined by finding the minimum of the new error. The process of locating new pulses is continued until the error is reduced to acceptable values or the number of pulses reaches the maximum value that can be encoded at the specified bit rate. The speech quality and the bit rate for the multi–pulse excitation are determined by the number of pulses; 4 to 8 pulses in a 5 msec frame are sufficient for producing high–quality speech.

7.2.4 HIGH–QUALITY SPEECH BELOW 8 kbits/sec

Recent speech coding work using stochastically–excited linear predictive coding has shown great promise for producing high–quality speech below 8 kbits/sec and possibly as low as 4.8 kbits/sec [16]. Such low rates are attractive for transmitting digital speech over narrow–band radio channels and for providing end–to–end digital speech communication over ordinary dial–up public telephone lines. The excitation in stochastic coders is determined by an exhaustive search from a codebook of white Gaussian sequences to minimize the perceptual distortion in the synthetic speech. The search procedure for stochastic excitation is illustrated in Fig. 5. These coders are extremely complex and require more than 50 million multiply/add operations per second. The rapid progress in custom VLSI circuits will enable us to handle this complexity in the next few years. The architecture of the stochastic coder is well suited for VLSI implementation since the search procedure carries out a large number of simple identical operations, namely, the computation of error for each member of the codebook.

STOCHASTIC CODING OF EXCITATION

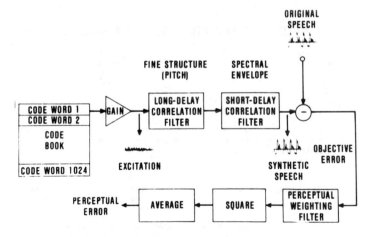

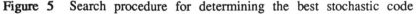

Figure 5 Search procedure for determining the best stochastic code

7.3   SPEECH SYNTHESIS

A principal objective in speech synthesis is to produce natural quality synthetic speech from unrestricted text input.   The goal is to provide great versatility for having a machine speak information to a human user, in as natural a manner as possible.   Useful applications of speech synthesis include announcement machines (e.g., weather, time), computer answer back (voice messages, prompts), information retrieval from databases (stock price quotations, bank balances), reading aids for the blind, and speaking aids for the vocally handicapped.

There are at least three major factors influencing the performance of speech synthesizers.   The first factor is the quality (or naturalness) of the synthesis.   It is often possible to trade between quality and message flexibility. For example, simple announcement machines often use the best speech coding methods to give high–quality speech, because the messages to be spoken are fixed in context and limited in number.   However, text–to–speech systems aim for great flexibility and message versatility, and, in this case, the speech signal must be synthesized from fundamental units.

A second factor is the size of the vocabulary.   If a relatively small vocabulary is required (e.g., 100–500 words), it is possible to custom–adjust the synthesis for improved naturalness.   However, for vocabularies of more than 1000 words, customized tuning is inappropriate.

A third factor affecting speech synthesis is the cost (or, complexity) of the system.   The cost includes hardware required for storage of words, phrases, dictionaries and production rules, as well as hardware required for speech signal generation (e.g., coder, synthesizer).   The cost of synthesis systems has fallen rapidly with advances in VLSI, so this factor is becoming less of an issue.

7.3.1   SPEECH SYNTHESIS FROM STORED CODED SPEECH

The easiest method of providing voice output for machines is to create speech messages by concatenation of prerecorded and digitally stored words, phrases, and sentences spoken by a human.   There are several trade–offs to be considered here.   Using words as the basic synthesis seem to be a proper choice, in many cases, because it allows one to create a large number of utterances from a relatively small number of words.   However, the process of joining words and creating sentences with the correct prosody is much more difficult.   This problem can be avoided by recording sentences, but the number of sentences to be stored increases exponentially with the number of words in a sentence.   In order to reduce the storage requirements, a variety of speech coding methods can be used.   Simple speech coding procedures can produce high–quality speech at 32 kbits/sec.   Speech coding techniques, such as multi–pulse LPC, can bring the data rate down to 10 kbits/sec.   At this bit rate, approximately 100 sec of speech data can be stored on a single 1 megabit ROM chip.   MPLPC is capable of producing high–quality speech using a simple speech synthesizer; most of the complexity of multi–pulse LPC is in the speech analysis part that has to be done only once and does not need real–time operation. Data rates as low as 1000 bits/sec can be realized using LPC vocoding techniques but the speech quality is much lower (the speech is intelligible but lacks naturalness) at these low data rates.

The flexibility of stored–speech synthesis systems can be further enhanced by allowing control of prosody (pitch and duration adjustments) during the synthesis process. The MPLPC technique is particularly suitable for providing the desired control of pitch and duration. With the decreasing cost of digital storage, stored–speech synthesis techniques could provide low cost voice output for many applications.

Figure 6 shows a block diagram of a general concatenative type of synthesis system. The storage consists of a fixed set of words, phrases, and sentences which have been encoded. An input message, which is a sequence of words, phrases, and sentences, is converted to the appropriate sequence of units which are retrieved and concatenated (usually with some type of smoothing at the junctions between units). The concatenated units are sent to a decoder (synthesizer) and to a digital–to–analog converter for transmission and/or playback. The concatenative type of synthesis is used primarily in announcement machines, and for applications such as automatic intercept of incorrectly dialed telephone numbers, where only a small vocabulary is required, and a limited set of output sentences is needed [17].

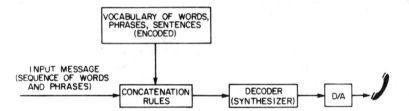

Figure 6 Block diagram of a concatenation type of speech synthesizer.

7.3.2 TEXT–TO–SPEECH SYNTHESIS

Stored–speech systems are not flexible enough to convert unrestricted printed text–to–speech – the objective of text–to–speech systems. Applications include accessing and speaking electronic mail, reading machines for the blind, and automated directory assistance systems that speak subscriber names and telephone numbers. A text–to–speech system must be able to accept the incoming text – which often includes abbreviations, Roman numerals, dates, times, formulas, and a wide variety of punctuation marks – and convert it into a speakable form. The text is translated into a phonetic transcription, using a large pronouncing dictionary supplemented by appropriate letter–to–sound rules. A stored library of about 2000 LPC (or formant) coded speech segments spans the range of speech sounds of a given language and provides the means for converting the phonetic elements to spoken form. Thus, the system is able to synthesize virtually unrestricted sequences of phonemes. Speech waveforms are finally generated from acoustic parameters using LPC or formant synthesis. The resulting speech is intelligible and acceptable for a variety of applications.

A block diagram of a text–to–speech synthesizer (TTS) is shown in Fig. 7. In its most general form, the input to the system is a message in the form of unrestricted ASCII text, and the output of the system is the continuously spoken message. The system has three major modules; letter–to–sound conversion, sound–to–parameter assembly, and synthesis from a parametric description of the text. The letter–to–sound conversion can utilize either a set of programmed pronouncing rules or a stored pronouncing dictionary (which provides the phonetic spelling of every word in the text message) or a mixture of these two techniques. Even with dictionaries of several hundred thousand words, there will be cases where the words of the ASCII text are not always found (e.g., proper names, cities, specialized terminology), and for such cases programmed pronunciation rules are mandatory. In addition to deriving the phonetic symbols that correspond to the text of the input message, the first module must also provide prosody markers (pitch, duration, intensity) for the message to be spoken.

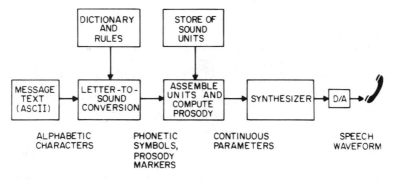
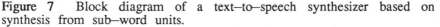

Figure 7 Block diagram of a text–to–speech synthesizer based on synthesis from sub–word units.

The second stage of the TTS system performs the conversion from phonetic symbols to continuous synthesis parameters, based on the set of sub–word units used to represent the speech. Thus if dyad units are used, a conversion from phonetic symbols to dyads is required, followed by retrieval and smoothing of the synthesis parameters corresponding to the dyads in the message. Continuous contours for pitch and timing are also computed in this stage. New work in automatic parsing and syntax analysis is providing improved capabilities in computing speech prosody. The final stage is the synthesis of speech from the parametric representation of the sub–words units. Typically, an LPC, MPLPC, or formant synthesizer is used [18,19].

TTS synthesizers can be used for database access, such as stock price quotations and bank balance checking, for access to voluminous amounts of text material over telephone lines (e.g., medical or legal encyclopedias), and as reading aids for the visually handicapped. Current TTS synthesizers produce speech which approaches the word intelligibility of natural speech, but the quality is typically synthetic sounding. They perform with large vocabularies and great flexibility and at relatively low cost. The challenge in speech synthesis over the next several years is to improve voice quality and increase flexibility by providing a range of voice styles (male, female, child), voice characteristics (Southern drawl, New England accent, etc), and different languages. In this manner, TTS systems can be tailored both for the application, and for the intended set of users.

Rapidly advancing VLSI technology will have a large impact on future speech synthesis technology. Present computer models of speech synthesis are simple in comparison to human speech generation and it is not yet practical to implement more sophisticated synthesis models. But future advances in fundamental understanding of speech production and language, and of syntactic and semantic analysis, will contribute significantly to improved text–to–speech synthesis.

7.4   SPEECH RECOGNITION

Figure 8 shows a block diagram of the traditional, pattern recognition based, speech recognition model. The input speech signal can be anything from a word (or a sequence of isolated words), to a sentence of continuous speech. The first processing block is feature measurement in which the speech signal is spectrally analyzed, periodically in time, to give a series of spectral feature vectors characteristic of the behavior of the speech signal. For the most part we have used linear predictive coding (LPC) as the spectral representation, but other spectral analyses like filter bank analysis are equally suitable [20]. The time sequence of spectral features is called a test pattern.

The second step in the processing is a pattern similarity measurement in which the running set of spectral vectors (the test pattern) is compared to a set of stored reference patterns, and for each such comparison a distance or similarity score results. For the most part we have used single words as the stored reference patterns, but even in cases in which the basic recognition units are smaller than words (e.g., syllables, demisyllables, dyads, phonemes, etc), a lexicon can be used to build word reference patterns and so we are

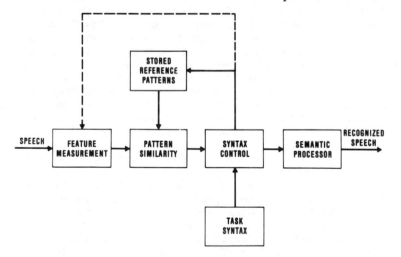

Figure 8   Block diagram of a speech recognizer incorporating syntax and semantics.

equivalently using words as the recognition unit. The pattern similarity measurement typically involves time registration of the stored reference pattern (which consists of a series of feature vectors) with the running speech (which is also a series of feature vectors). The technique of dynamic time warping (DTW) is generally used to provide the optimal alignment between references and test (speech) patterns [21].

The basic procedures of time alignment are illustrated in Fig. 9 which shows representative contours of a test and reference pattern (the lengths of both patterns have been made equal here; in general they are different and this difference must be accounted for). It can be seen that distinctive events in the two patterns (i.e., peaks in the contours) do not occur at the same time instants. Thus, the purpose of DTW alignment procedure is to derive an optimal time alignment between test and reference patterns by locally shrinking or expanding the time axis of one of the patterns to optimally match the other pattern. An efficient mathematical procedure exists for obtaining an optimal alignment curve based on dynamic programming techniques [21]. The alignment curve for the examples of Fig. 9 is shown at the upper right of this figure. The similarity (or equivalently distance) between a reference and test pattern is defined as the normalized sum of the spectral similarities (distances), along the discrete set of point in the optimal time alignment path, between reference and test patterns.

The third step in the processing of Fig. 8 is syntax control which uses task syntax to determine the proper sequencing of stored reference patterns (words) for the task at hand. The syntax control could, in theory, also exercise control over the feature measurement algorithm, thereby changing the type (and/or form) of analysis depending on the sound to be recognized. Such sophisticated control has not been used in current speech recognizers.

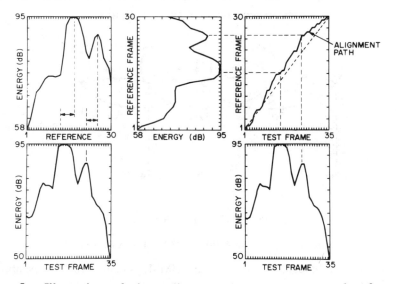

Figure 9   Illustration of time alignment between a test and reference pattern.

The last step in processing of Fig. 8 is a semantic processor which chooses, as the recognized speech, the sentence (or word) which has both the smallest distance (or highest similarity) to the input speech, and which is semantically meaningful (given that it has already been checked for syntax).

7.4.1   HMM MODELS

An alternative to using templates to characterize words (or sub–word units) is to build probabilistic models which describe, statistically, the time–varying spectral characteristics of the word. One very popular form of such probabilistic models is the hidden Markov model (HMM) [7], an example of which is shown in Fig. 10. This model has N states (5 in the example shown) and each state physically corresponds (in some vague sense) to a set of temporal events in the speech sound. The overall HMM is characterized by a state transition matrix, A, (which describes how new states may be reached from old states) and by a statistical characterization of the acoustic vectors, B (the analysis feature vectors, x), within the state.

The changes, to the recognition structure of Fig. 8, required because of using HMM's rather than templates are minimal. The store of template reference patterns is replaced by a store of reference models, and the pattern similarity algorithm uses statistical scoring instead of distances and uses a somewhat different alignment procedure to line up states of the reference model to frames of the test pattern.

7.4.2   PERFORMANCE RESULTS – ISOLATED WORDS

For isolated word recognition, the classic technique has been to build templates or statistical models based on natural spoken occurrences of the word. In the simplest case, a word reference pattern is created directly from

HIDDEN MARKOV MODEL (LEFT-TO-RIGHT)

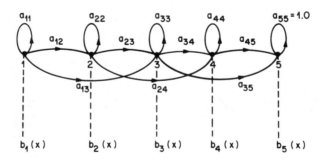

$$A = [a_{ij}] = PROB\,(STATE\ j\ |\ STATE\ i)$$

$$B = [b_j\,(x)] = PROB\,(ANALYSIS\ VECTOR\ x\ |\ STATE\ j)$$

Figure 10   A hidden Markov model (HMM) suitable for representing a single word.

Table 1   Performance of isolated word recognizers as a function of vocabulary size.

Vocabulary	Speaker Mode	Accuracy
10 digits	SI	99.2%
39 Alphadigits	SD	95.5%
	SI	89.5%
54 Computer Terms	SI	96.5%
129 Airline Terms	SD	88%
	SI	91%
1109 Basic English	SD	79.2%

one or more spoken occurrences of the word by a given talker.   In a more sophisticated application, a set of multiple occurrences of the word is clustered to give one (or more) word reference patterns.   The patterns may be talker specific (the so–called speaker–dependent (SD) recognizers), or speaker–independent (SI), depending on the way they are derived.   The vocabulary sizes, for which isolated word systems have been tested, range from a few words (e.g., 10 digits), up to over 1000 words, (e.g., 1109 words of Basic English).   Table 1 summarizes the current performance for a range of vocabularies, for both SD and SI cases.   It can readily be seen that the complexity of the words in the vocabulary (i.e., how similar are the nearest sounding word pairs) is more important than mere vocabulary size.

7.4.3   CONNECTED WORD RECOGNITION

A somewhat more complicated task in speech recognition is that of recognizing speech which is nominally spoken as a connected word string, e.g., digit strings for dialing telephone numbers, letter strings for spelling names, etc.   The manner in which such strings are recognized, based on the statistical pattern recognition approach, is illustrated in Fig. 11.   We assume each word in the vocabulary is represented by one or more reference patterns (i.e., templates or statistical models) and that the unknown spoken word string can be recognized by finding the best concatenation of reference patterns which matches the test pattern.   There are several problems associated with trying to find the optimal sequence of reference patterns to match the unknown test pattern, including:

1.   The number of words in the test pattern is generally unknown.

2.   The locations, in time, of the boundaries between words is unknown; in fact there are really no well defined boundaries in many cases since the end of one word often merges smoothly with the beginning of the following word.

3.   Matches between reference and test patterns are generally poor at the beginnings and ends of reference patterns because of the high degree of variability.

4.   Combinations of matching strings exhaustively (i.e., by trying all combinations, of all lengths, of all reference patterns) is prohibitively expensive.

Fortunately, several algorithms have been devised which optimally solve the matching problem without an exponential growth in computation as the vocabulary or size of the string grows [22–25]. One such algorithm is the level building (LB) procedure, which allows the recognition processing to proceed in a series of levels (words) to determine the best connected word string match for every permissible string length. Thus solutions to problems 1 and 4, above, have been found. No perfect solution to problems 2 and 3 is known. However, a reasonable approach, and one which has worked quite well to date, is to extract word reference patterns from tokens obtained from connected word strings. Thus the reference patterns for digits, for example, are obtained from analysis of a training set of connected digit strings; hence each reference pattern has information about the spectral dynamics of digits in strings, rather than in isolation.

CONNECTED WORD RECOGNITION FROM WORD TEMPLATES

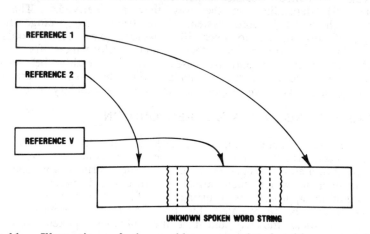

Figure 11   Illustration of the problems associated with recognizing a connected word string from single word reference patterns.

7.4.4   PERFORMANCE RESULTS – CONNECTED WORDS

Table 2 summarizes current performance of connected word recognizers based on the LB algorithm. For a digits vocabulary, in a speaker trained mode, string accuracies greater than 99% have been obtained for unknown length strings. In a speaker–independent mode, the best string accuracy has been 97% for unknown length strings and 98% for known length strings. Results are also given in Table 2 for connected letter recognition of spelled names from a 17,000 name directory, and for an airlines reservation and information task based on a vocabulary of 127 words.

Table 2 Performance of connected word recognizers.

Vocabulary	Speaker Mode	Word Accuracy	Task	String (Task) Accuracy
10 digits	SD	>99%	Random Strings	>99% UL
				>99% KL
	SI	97.5%	1–7 Digits	>99% UL
				>99% KL
26 Letters	SD	≃ 80%	Directory Listing	96%
	SI	≃ 80%	Retrieval 17,000 Names	90%
127 Airline Terms	SD	96%	Airlines Reservation and Information	97%
	SI	93%		75%

UL is for Unknown Length Strings
KL if for Known Length Strings

7.4.5 CONTINUOUS SPEECH RECOGNITION

Based on experience with more limited speech recognition tasks, work has begun on building a large vocabulary (1000–2000 word), natural syntax (i.e., approaching that of spoken English) continuous speech recognition system. A block diagram of the proposed system architecture is given in Fig. 12. This similarity of Fig. 12 to Fig. 8 should be obvious to the reader. The major complications in building such a recognizer are the following:

1. Words cannot be the basic unit for recognition; instead sub–word units must be used. Possible sub–word units include syllables, demi–syllables, diphones, dyads, phonemes, etc.

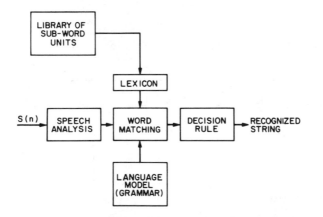

Figure 12 Model for large vocabulary speech recognition based on sub–word speech units.

2.   A lexicon must be used which describes how words are made up from the sub–word units. The lexicon can be an explicit representation (e.g., a dictionary of pronunciations from sub–word units), or it can be probabilistic in nature.

3.   A language model is used to describe the constraints among words in the language. The language model could be a formal grammar, a statistical model, or even an explicit state diagram of task syntax as used in Fig. 8.

Each of the complications listed above is formidable and leads to a wide range of choices of how to handle the problem. Taken together, they give an idea as to why continuous speech recognition is, and will remain, an unsolved problem for a long time.

7.5   SPEAKER RECOGNITION

The speaker recognition problem is really a pair of problems – namely, speaker identification, in which a talker is identified as one of a given set of talkers, and speaker verification, in which the talker gives both a claimed identity, and a transaction request, and the system decides whether to accept or reject the identity claim. It should be clear that speaker identification is a much harder problem than speaker verification, since, as the number of speakers increase without bound, the probability of error goes to 1 in identifying a talker, whereas the probability of error remains fairly constant for speaker verification.

Figure 13 shows a block diagram of a speaker recognition system. The input speech, which can be either a sentence, or a sequence of words (e.g., digits), is first spectrally analyzed, and then the resulting spectral pattern

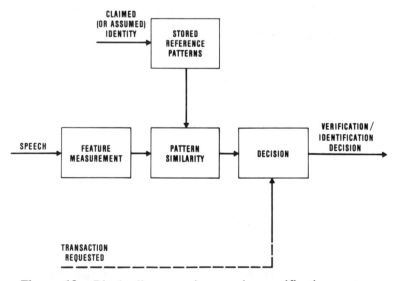

Figure 13   Block diagram of a speaker verification system.

is compared to stored reference patterns, using DTW methods. For speaker identification, the pattern similarity processing must be performed for each assumed talker (i.e., for the entire set of talkers), and the decision box chooses the identified talker as the one with the highest similarity to the input speech. For speaker verification, the pattern similarity processing is only performed for the claimed identity, i.e., only a single distance score results. Based on the transaction requested and the similarity score of the DTW processing, the decision box decides whether to accept or reject the claimed identity. Thus, for a banking transaction, a lower degree of similarity would be required to deposit money into an account, than to withdraw money from the account.

It can be seen, by comparing Figs. 8 and 13, that the processing for speech and speaker recognition is quite similar. Thus, as fundamental improvements are made in any of the basic procedures (feature measurement, pattern similarity, etc.), the performance of both types of systems improves.

Key factors affecting the performances of speaker verification systems are the type of input string which is used, the features used to characterize the voice pattern, and the type of transmission system over which the verification system is used. Best performance is achieved when sentence long utterances are used in a relatively noise–free speaking environment. Conversely, poorer performance is achieved for short, unconstrained, spoken utterances,in a noisy environment. Table 3 summarizes current performance of several types of speaker verification systems [26–27]. Current research in speaker recognition aims to improve performance by adapting the talker patterns, over time, to track changes in voice patterns.

7.6 CONCLUDING COMMENT

This overview of digital speech processing has aimed to highlight recent advances, current areas of research, and key issues for which new fundamental

Input	Mode	Acoustic Signal Processing	Feature Pattern	Verification Performance (Equal-Error Rate)
Sentence-Long Utterances	Text Dependent	10th Order Cepstral Analysis	Time Contours of Cepstral Coefficients	1% Recorded-Telephone 4% Live-Telephone
Isolated Word Strings	Text Independent	8th Order LPC Analysis	Speaker-Dependent Word Templates Speaker Independent Template Distance	4% Recorded-Telephone 8% Recorded-Telephone
Isolated Word Strings	Text Independent	8th Order Cepstral Analysis	Vector Quantization Codebook Talker Models	1% Recorded-Telephone

Table 3   Performance of speaker verification system.

understanding of speech is needed. Future progress in speech processing will surely be linked closely with advances in computation, microelectronics and algorithm design.

7.7 REFERENCES

1.   "Mini–supercomputer Boasts Integrated Approach to Vector Processing", Computer Design, p. 109, Sept. 1983.

2.   R.N. Kershaw et al., "A Programmable Digital Signal Processor with 32B Floating Point Arithmetic", Proc. ISSC, pp. 90–91, Feb. 1985.

3.   B.S. Atal and J.R. Remde, "A New Model of LPC Excitation for Producing Natural–Sounding Speech at Low Bit Rates", Proc. ICASSP–82, pp. 614–617, Paris, France, April 1982.

4.   M.R. Schroeder and B.S. Atal, "Code–Excited Linear Prediction (CELP): High Quality Speech at Very Low Bit Rates", Proc. ICASSP–85, Paper 25.1, pp. 937–940, March 1985.

5.   C.S. Meyers and L.R. Rabiner, "A Level Building Dynamic Time Warping Algorithm for Connected Word Recognition", IEEE Trans. on Acoustics, Speech and Signal Processing, Vol. ASSP–29, No. 2, pp. 284–297, April 1981.

6.   F. Jelinek, "Continuous Speech Recognition by Statistical Methods", Proc. IEEE, Vol. 64, No. 4, pp. 532–556, April 1976.

7.   S.E. Levinson, L.R. Rabiner and M.M. Sondhi, "An Introduction to the Application of the Theory of Probabilistic Functions of a Markov Process to Automatic Speech Recognition", Bell Systems Tech. J., Vol. 62, No. 4, pp. 1035–1074, April 1983.

8.   N.S. Jayant, "Coding Speech at Low Bit Rates", IEEE Spectrum, Vol. 23, No. 8, pp. 58–63, August 1986.

9.   W.P. Hayes et al., "A 32–bit VLSI Digital Signal Processor", IEEE Jour. Solid–State Circuits, Vol. SC–20, No. 5, pp. 998–1004, Oct. 1985.

10.   H. Alrutz, "Implementation of a Multi–Pulse Coder on a Single Chip Floating–Point Processor", Proc. 1986 IEEE Int. Conf. on Acoustics, Speech, and Signal Proc., Tokyo, Japan, pp. 2367–2370, April 1986.

11.   B.S. Atal, "Predictive Coding of Speech at Low Bit Rates", IEEE Trans. Commun., Vol. COM–30, pp. 600–614, April 1982.

12.   F.K. Soong, R.V. Cox and N.S. Jayant, "A High Quality Subband Speech Coder With Backward Adaptive Predictor and Optimal Time–Frequency Bit Assignment", Proc. 1986 IEEE Int. Conf. on

Acoustics, Speech, and Signal Proc., Tokyo, Japan, pp. 2387–2390, April 1986.

13.   S. Singhal and B.S. Atal, "Improving Performance of Multi–Pulse LPC Coders at Low Bit Rates", in Proc. 1984 IEEE Int. Conf. on Acoustics, Speech, and Signal Proc., Vol. 1, Paper No. 1.3, March 1984.

14.   T. Tremain, Personal communication 1985.

15.   B.S. Atal and J.R. Remde, "A New Model of LPC Excitation for Producing Natural–Sounding Speech at Low Bit Rates", Proc. 1982 IEEE Int. Conf. on Acoustics, Speech, and Signal Processing, Paris, France, pp. 614–617, 1982.

16.   B.S. Atal and M.R. Schroeder, "Stochastic Coding of Speech Signals at Very Low Bit Rates", Proc. Int. Conf. Commun. – ICC84, Part 2, pp. 1610–1613, May 1984.

17.   L.R. Rabiner and R.W. Schafer, "Digital Techniques for Computer Voice Response: Implementations and Applications", Proc. IEEE, Vol. 64, No. 4, pp. 416–433, April 1976.

18.   J. Allen, "Synthesis of Speech From Unrestricted Text", Proc. IEEE, Vol. 64, pp. 433–442, April 1976.

19.   J.P. Olive, "A Scheme for Concatenating Units for Speech Synthesis", Proc. ICASSP–80, pp. 568–571, Denver, Colorado, April 1980.

20.   J.D. Markel and A.H. Gray Jr., Linear Prediction of Speech, New York: Springer–Verlag, 1976.

21.   F. Itakura, "Minimum Prediction Residual Principle Applied to Speech Recognition", IEEE Trans. on Acoustics, Speech, and Signal Processing, Vol. ASSP–23, No. 1, pp. 66–72, Feb. 1975.

22.   H. Sakoe, "Two Level DP Matching – A Dynamic Programming Based Pattern Matching Algorithm for Connected Word Recognition", IEEE Trans. on Acoustics, Speech, and Signal Processing, Vol. ASSP–27, pp. 588–595, Dec. 1979.

23.   C.S. Myers and L.R. Rabiner, "Connected Digit Recognition Using a Level Building DTW Algorithm", IEEE Trans. on Acoustics, Speech, and Signal Processing, Vol. ASSP–29, No. 3, pp. 351–363, June 1981.

24.   J.S. Bridle, M.D. Brown and R.M. Chamberlain, "An Algorithm for Connected Word Recognition", Automatic Speech Analysis and Recognition, J.P. Haton, Ed., pp. 191–204, 1982.

25.   J.L. Gauvain and J. Mariani, "A Method for Connected Word Recognition and Word Spotting on a Microprocessor", Proc. 1982 ICASSP, pp. 891–894, May 1982.

26.   S. Furui, "Cepstrum Analysis Technique for Automatic Speaker Verification", IEEE Trans. on Acoustics, Speech, and Signal Processing, Vol. ASSP–29, No. 2, pp. 254–272, April 1981.

27.   F.K. Soong, A.E. Rosenberg, L.R. Rabiner and B.H. Juange, "A Vector Quantization Approach to Speaker Recognition", Proc. ICASSP '85, pp. 387–390, April 1985.

8
IMAGE PROCESSING AND ANALYSIS USING PRIMITIVE COMPUTATIONAL ELEMENTS

WILLIAM K. PRATT
Vicom Systems, Inc.
San Jose, CA

8.0   ABSTRACT

It is possible to decimate most image processing and analysis algorithms into series and parallel interconnections of primitive computational elements. These primitive elements include: a point processor, an ensemble processor, a local neighborhood spatial processor, a pixel integrator, a histogram generator, and a feature list generator. In a pipeline image processor, these primitive computational elements can be realized by relatively simple electronic circuits, or even single integrated circuit chips, that operate at digital video rates.

This paper describes primitive computational element decompositions of image processing and analysis algorithms.

8.1   INTRODUCTION

Image processing and analysis, by nature, involve two dimensional operations on an image. In image processing, an input image is transformed to produce an output image. Applications include transformations: to reduce the effects of image noise or interference, to enhance the sharpness of image detail, and to minify or magnify the size of an image. Image analysis operations ultimately involve transformations on an image to produce a concise description of the image; for example, to identify objects within an image. Typically, however, there are many preliminary steps in an image analysis algorithm, leading to the final description, in which an image is spatially processed to detect features, such as edges or texture.

An indepth review of image processing and analysis algorithms indicates that they can be decimated into a relatively few primitive computational elements that can be connected in serial or parallel [1]. A set of primitive image processing computational elements is defined below:

Point Processor
> A pixel–by–pixel linear or nonlinear modification of an input pixel to produce an output pixel.
> Example: magnitude, logarithm, square.

Ensemble Processor
> A linear or nonlinear combination of a pair of input pixels or of an input pixel and a constant to produce an output pixel.
> Example: addition, multiplication, exclusive OR, maximum.

Spatial Processor
A linear or nonlinear combination of input pixels within some neighborhood to produce an output pixel.
Example: convolution, morphological erosion.

A set of primitive image analysis computational elements is listed below:

Pixel Integrator
A sum of all processed pixels within an image.

Histogram Generator
A count of the amplitude level occurrence of pixels within an image.

Feature List Generator
A list of the row and column addresses and pixel amplitudes of pixels meeting some amplitude criterion.

The motivation for the decimation of a complex image processing or analysis algorithm into series or parallel interconnected primitive computational elements is primarily a desire to reduce implementation complexity and to improve the rate of computation in comparison with a general purpose computer. With primitive computational elements, it is possible to develop relatively simple circuits or even single integrated circuit chips that can execute the primitive operations at digital video rates using a recirculating, serial or parallel pipeline structure.

Figure 1 provides an example [1] of the architecture of a Pipeline Image Processor (PIP). This processor combines the primitive processing computational elements in a manner in which they are commonly joined together in an algorithm flow chart. In the PIP, the data paths are all 16 bits in width. The spatial processors perform convolution, maximum and minimum functions within a 3 x 3 pixel window. The point processors perform 65,536 entry lookup table (LUT) transformations. The ensemble processors perform 70 programmable combinations of a pair of pixels or a pixel and a constant.

Figure 2 illustrates the concept of a recirculating pipeline structure, which employs a single Pipeline Image Processor and a single Pipeline Image Analyzer. In this structure, an image is recirculated from image memory through a PIP a number of times. In the Pipeline Image Analyzer, the three primitive analysis computational elements passively monitor, in parallel, the

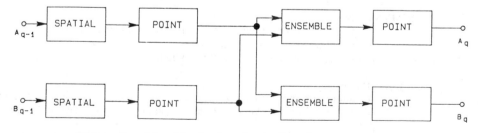

Figure 1 Algorithmic–based pipeline image processor.

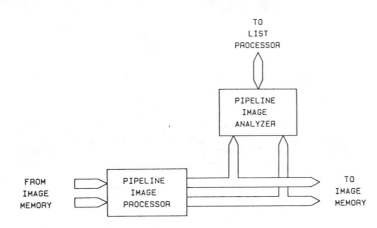

Figure 2   Recirculating pipeline connection.

output image data bus from the PIP and continuously compute feature lists. These lists are subsequently manipulated by a general purpose computer or by a list processor.   After each recirculation cycle, the PIP and PIA are reprogrammed to perform the next step of the algorithm. The disadvantage of this configuration is the frame time delay incurred during each recirculation. Figure 3 describes a serial pipeline structure.   The pixel processing rate is the same as the input pixel rate.   More processing elements can be added serially as the complexity of an algorithm increases but the pipeline processing rate remains constant.   Of course, some time latency is incurred in filling the pipeline.   A parallel connection of PIPs and PIAs is shown in Figure 4.   In this structure, several different processing and analysis algorithms can be executed simultaneously.   Compound parallel and serial computing structures can also be developed by combining the architectures of Figures 3 and 4.

This paper concentrates on decimation algorithms rather than specific implementations.   It will be shown that for some algorithms, such as two-dimensional convolution, it is possible to perform an algorithmic decimation into primitive elements in a systematic analytic manner.   For other algorithms, there is no existing proof of such a decimation, nor are there any automatic

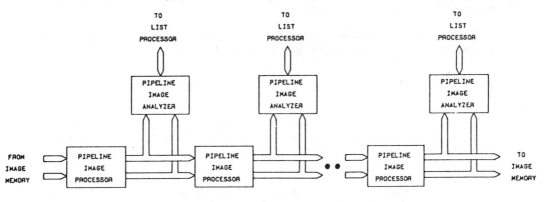

Figure 3   Serial pipeline connection.

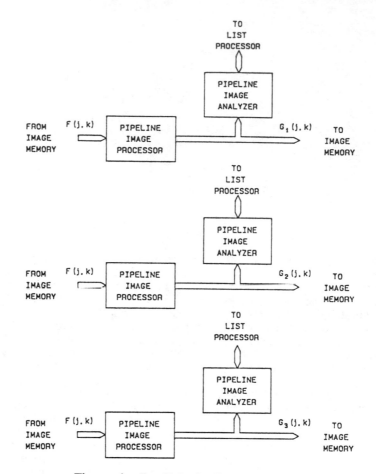

Figure 4   Parallel pipeline connection.

decompositions.   In such instances, the only recourse is case–by–case cleverness.

The following sections illustrate the primitive element decomposition process for several image processing and analysis algorithms.

8.2   SPATIAL FILTERING

Spatial filtering is a widely used technique in image processing for noise reduction, detail sharpening and a variety of other applications. It can be implemented by a convolution operation in which an output image is formed by a linear combination of pixels of an input image within some local neighborhood. The convolution operation is defined by

$$G(j,k) \quad = \quad \sum_{m} \sum_{n} F(m,n) \ H[j-m+C, \ k-n+C] \qquad (1a)$$

or symbolically as

$$G(j,k) = F(j,k) \circledast H(j,k) \tag{1b}$$

where $F(m,n)$ and $G(j,k)$ are input and output images, respectively, that are assumed square over the indices $1 \leq m,n,j,k \leq N$. The term $H(j,k)$ is the impulse response array for $1 \leq j,k \leq L$. It contains the spatial weighting values of the convolution operation. The limits of the summation are $j-Q \leq m \leq j+Q$ and $k-Q \leq n \leq k+Q$, where $Q = (L-1)/2$. Equation 1 defines the output image over the limits $C \leq j,k \leq N-C+1$, where $C = (L+1)/2$; the remaining border elements in the $N \times N$ output image are set to zero.

In the special case for which $L = 3$, the convolution operation can be written explicitly as

$$\begin{aligned}
G(j,k) = \; &H(3,3)F(j-1,k-1) + H(3,2)F(j-1,k) + H(3,1)F(j-1,k+1) \\
&+ H(2,3)F(j,k-1) + H(2,2)F(j,k) \quad + H(2,1)F(j,k+1) \\
&+ H(1,3)F(j+1,k-1) + H(1,2)F(j+1,k) + H(1,1)F(j+1,k+1)
\end{aligned} \tag{2}$$

for $2 \leq j,k \leq N-1$. Figure 5 lists several common 3 x 3 pixel impulse response arrays for spatial filtering applications.

```
   LOW PASS        NEIGHBOR        HIGH PASS
    FILTER          AVERAGE          FILTER


  0   0   0        1   1   1       0  -1   0

  0   1   1        1   0   1      -1   5  -1

  0   1   1        1   1   1       0  -1   0


  1   1   1        0   1   0      -1  -1  -1

  1   1   1        1   0   1      -1   9  -1

  1   1   1        0   1   0      -1  -1  -1


  1   1   1        1   0   1       1  -2   1

  1   2   1        0   0   0      -2   5  -2

  1   1   1        1   0   1       1  -2   1


  1   2   1

  2   4   2

  1   2   1
```

Figure 5   SGK Filter kernels.

8.2.1   SGK CONVOLUTION

The impulse response array H(j,k) of Eq. 1 can be decimated into a sequence of 3 x 3 pixel arrays, each of which is called a Small Generating Kernel (SGK) [2,3]. The expansion, which generally is approximate, is given by

$$\hat{H}(j,k) = K_1(j,k) \circledast \cdots \circledast K_q(j,k) \circledast \cdots \circledast K_Q(j,k) \tag{3}$$

where $\hat{H}(j,k)$ is an approximation to H(j,k) and $K_q(j,k)$ is the q–th SGK of the decomposition with Q = (L–1)/2. Techniques, which have been developed for choosing the SGK's to minimize the mean square error between $\hat{H}(j,k)$ and H(j,k), result in an error of 5% or less [2].

8.2.2   SVD/SGK CONVOLUTION

An arbitrary impulse response array H(j,k) can be exactly factored into a sequence of SGKs by use of the Singular Value Decomposition (SVD) technique combined with Small Generating Kernel convolution [4,5]. Let the prototype impulse reponse array H(j,k) be represented as a matrix H. It is then possible to express H as a sum of unit rank matrices by the SVD expansion

$$H = \sum_{i=1}^{R} H_i = \sum_{i=1}^{R} S_i \, a_i \, b_i \tag{4}$$

where R ≤ L is the rank of H, S_i is the i–th singular value of H, and a_i and b_i are the eigenvectors of HH^T and H^TH, respectively. Each constituent matrix H_i can then be subjected to an SGK expansion, without approximation error, to obtain the decomposition

$$H_i(j,k) = K_{i1}(j,k) \circledast \cdots \circledast K_{iq}(j,k) \circledast \cdots \circledast K_{iQ}(j,k) \tag{5}$$

where $K_{iq}(j,k)$ is the q–th SGK corresponding to the i–th singular value. Figure 6 contains a flow chart of the SVD/SGK algorithm. Typically, the amplitude–ordered singular values S_i diminish in amplitude quite rapidly with respect to the index, and consequently, the summation can be truncated with little error.

8.3   INTERPOLATION

In image processing display systems, it is often necessary to "blow–up" or "zoom" an image to better view a sub–area of the image. This necessitates the interpolation of pixels to produce a "smooth" display [6, p. 113]. The simplest form of interpolation is to repeat intermediate pixels between data

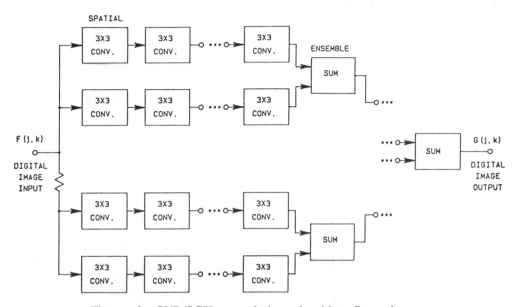

Figure 6   SVD/SGK convolution algorithm flow chart.

values. A smoother result can be obtained by linearly interpolating data values along rows and columns of the original image.   Higher–order interpolation produces even smoother images.

As an example, consider a N x N pixel image F(j,k), which is to be interpolated to size 2N x 2N.   Let F(j,k) be spatially "stretched" such that pixels of F(j,k) are stored in an array G(j,k), where every other row and column is set to zero.   The next step of the interpolation process is convolution in which G(j,k) is convolved with an interpolation operator.   Figure 7 lists the elements of several common interpolation operators.   Each operator should be scaled before convolution so that the sum of the elements equals four.

The square function operator provides interpolation by pixel replication. The triangle interpolation function results in first–order linear interpolation.   It should be noted that triangle interpolation can be achieved by successive square interpolation on the stretched image.   Convolving the stretched image three times successively with the square function is equivalent to bell function interpolation.   Cubic B–spline interpolation can be achieved by convolving the triangle operator with itself, or equivalently, convolving the square function with itself four times.   The concept described above can be readily extended for higher order zoom factors.

8.4   ENHANCEMENT

Image enhancement involves the modification of an image to render it more subjectively pleasing.   Typical operations include contrast improvement, noise attenuation and edge sharpening.

SQUARE:

```
1  1

1  1
```

TRIANGLE:

```
1   2   1

2   4   2

1   2   1
```

BELL:

```
1   3   3   1

3   9   9   3

3   9   9   3

1   3   3   1
```

CUBIC B-SPLINE:

```
1   4    6    4   1

4  16   24   16   4

6  24   36   24   6

4  16   24   16   4

1   4    6    4   1
```

Figure 7   Interpolation function arrays.

In many applications, contrast improvement can usually be accomplished by simple point processing. But, in some cases, this method is inadequate. A common instance is the enhancement of detail within shadowed regions of an image. This can be accomplished by the Wallis statistical differencing algorithm in which each pixel is "normalized" by subtraction of its neighborhood mean and division by its neighborhood standard deviation [7]. The "moving window" mean and standard deviation can be computed by SGK methods. Figure 8 illustrates the computation of the mean and standard deviation using SGK techniques. Figure 9 shows an example of the Wallis algorithm.

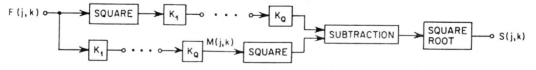

Figure 8   Mean and standard deviation computation.

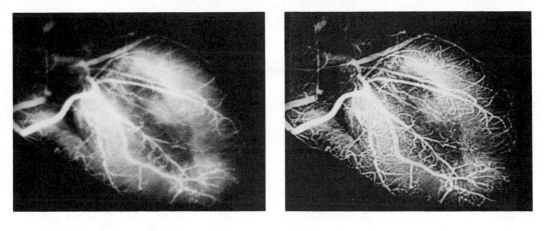

(a)   Original   (b)   Processed

Figure 9   Example of Wallis statistical differencing processing.

8.5   FEATURE EXTRACTION

Feature extraction algorithms are especially amenable to decimation into primitive processing and analysis computational elements. The following contains examples of spot and line detection, edge detection and texture discrimination [6, p. 471].

8.5.1   SPOT AND LINE DETECTION

Spots consisting of a single high amplitude pixel with a lower amplitude background can be detected by convolving an input image with one of the discrete Laplacian operators listed in Fig. 10, and then thresholding the resultant output. Similarly, lines oriented in 45 degree increments can be detected using the line detector kernels of Fig. 10. Spots and lines of larger dimension can be detected by spatially repeating each entry of the kernels in Fig. 10. Then the SVD/SGK technique can be applied to the large arrays to obtain 3 x 3 SGK kernels for implementation.

8.5.2   EDGE DETECTION

Figure 11 contains a flow chart for a form of edge detection based on edge gradient thresholding. In the algorithm, an input image is separably differentiated along rows and columns by a pair of differential kernels. After the row G_R and column G_C gradients are formed, they are merged and fed to a look–up table. The table is programmed to generate an output pixel $G(j,k)$, which is the edge orientation specified by an integer 1 to 8 corresponding to the eight compass directions of the edge gradient. If the edge magnitude is less than a threshold value, $G(j,k)$ is set to zero. Subsequent spatial analysis can be performed to delete false edges caused by noise, which will usually have opposing orientations.

SPOT			UNWEIGHTED LINE			WEIGHTED LINE		
0	-1	0	-1	-1	-1	-1	-2	-1
-1	4	-1	2	2	2	2	4	2
0	-1	0	-1	-1	-1	-1	-2	-1
-1	-1	-1	-1	2	-1	-1	2	-1
-1	8	-1	-1	2	-1	-2	4	-2
-1	-1	-1	-1	2	-1	-1	2	-1
1	-2	1	-1	-1	2	-2	-1	2
-2	4	-2	-1	2	-1	-1	4	-1
1	-2	1	2	-1	-1	2	-1	-2
			2	-1	-1	2	-1	-2
			-1	2	-1	-1	4	-1
			-1	-1	2	-2	-1	2

Figure 10   Spot and line detector kernels.

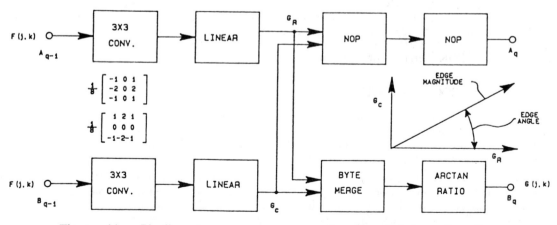

Figure 11   Pipeline image processor template for Sobel angle edge detection algorithm.

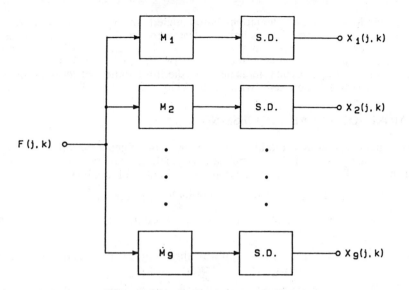

Figure 12 Laws texture masks.

8.5.3 TEXTURE DISCRIMINATION

Laws [8] has developed a simple, but highly effective method of image texture discrimination based on primitive processing computational elements. The flow chart is shown in Figure 12. In operation, an image is convolved with each of the nine texture detection masks defined in Fig. 13, which are designed to accentuate micro–structure in the image. Next, the

```
 1  2  1        -1  0  1        -1  2 -1
 2  4  2        -2  0  2        -2  4 -2
 1  2  1        -1  0  1        -1  2 -1


 1  2  1         1  0 -1         1 -2  1
 0  0  0         0  0  0         0  0  0
-1 -2 -1        -1  0  1        -1  2 -1


-1 -2 -1         1  0 -1         1 -2  1
 2  4  2        -2  0  2        -2  4 -2
-1 -2 -1         1  0 -1         1 -2  1
```

Figure 13 Laws texture features.

moving window standard deviation is computed in a P x P window on each convolved image using the technique described in Fig. 8. The size of the window is chosen to encompass several spatial periods of the texture field. The result is a nine element feature vector at each pixel that describes the image texture. Experiments indicate that the probability of correct classification typically exceeds 90 per cent with this technique.

8.6 MORPHOLOGICAL PROCESSING

Morphological operations modify the structure of objects within an image as a means of analysis [9,10]. Applications include defect detection in printed circuit boards, fingerprint identification and blood cell analysis.

There are three fundamental morphological operations:

1. Dilation — uniform growing of the spatial extent of an object.

2. Erosion — uniform shrinking of the spatial extent of an object.

3. Skeletonization — stick figure representation of an object.

These operations, and more complex variations of the operations, can be implemented by sequential processing in 3 x 3 pixel windows. Although techniques exist for both binary and grey scale images, only the former case is discussed here.

Consider a binary–valued N x N input image F(j,k), and a binary valued L x L "structuring element" H(j,k). The generalized dilation operation is defined as

$$G(j,k) \;\; = \;\; \underset{m,n}{\cup \cup} \; F(m,n) \; \cap \; H[j-m+C, \; k-n+C] \tag{6a}$$

or symbolically as

$$G(j,k) \;\; = \;\; F(j,k) \;\; \oplus \;\; H(j,k) \tag{6b}$$

where $\oplus$ denotes generalized dilation, $\cup$ denotes the union set operation (OR), $\cap$ denotes the intersection set operation (AND) and $C = (L + 1)/2$.

The limit ranges are the same as for convolution. If $L = 3$, then the generalized dilation operation can be written explicity as

$$G(j,k) = [H(3,3) \cap F(j-1,k-1)] \cup [H(3,2) \cap F(j-1,k)] \cup [H(3,1) \cap F(j-1,k+1)]$$
$$\cup \; [H(2,3) \cap F(j,k-1)] \cup [H(2,2) \cap F(j,k)] \cup [H(2,1) \cap F(j,k+1)]$$
$$\cup \; [H(1,3) \cap F(j+1,k-1)] \cup [H(1,2) \cap F(j+1,k)] \cup [H(1,1) \cap F(j+1,K+1)]$$

$$\tag{7}$$

for $2 \le j,k \le N-1$. Generalized erosion is defined as

$$G(j,k) \;\; = \;\; \underset{m,n}{\cap \cap} \; F(m,n) \; \cup \; \overline{H}[j-m+C, \; k-n+C] \tag{8a}$$

or symbolically as

$$G(j,k) = F(j,k) \ominus H(j,k) \tag{8b}$$

where $\ominus$ denotes generalized erosion and the overbar denotes the logical complement. For L = 3, generalized erosion can be written as

$$G(j,k) = [\overline{H}(3,3) \cup F(j-1,k-1)] \cap [\overline{H}(3,2) \cup F(j-1,k)] \cap [\overline{H}(3,1) \cup F(j-1,k+1)]$$

$$\cap [\overline{H}(2,3) \cup F(j,k-1)] \cap [\overline{H}(2,2) \cup F(j,k)] \cap [\overline{H}(2,1) \cup F(j,k+1)]$$

$$\cap [\overline{H}(1,3) \cup F(j+1,k-1)] \cap [\overline{H}(1,2) \cup F(j+1,k)] \cap [\overline{H}(1,1) \cup F(j+1,k+1)] \tag{9}$$

Figure 14 contains examples of dilation and erosion.

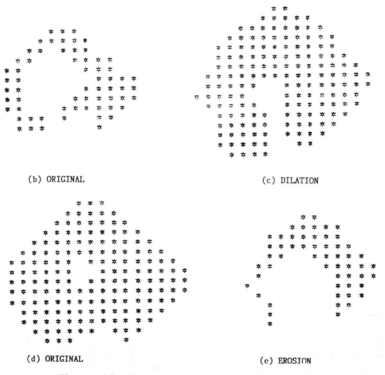

Figure 14   Examples of dilation and erosion.

It has been established that

$$[F(j,k) \oplus A(j,k)] \oplus B(j,k) = F(j,k) \oplus [A(j,k) \oplus B(j,k)] \qquad (10a)$$

and

$$[F(j,k) \ominus A(j,k)] \ominus B(j,k) = F(j,k) \ominus [A(j,k) \oplus B(j,k)] \qquad (10b)$$

where $A(j,k)$ and $B(j,k)$ are structuring elements. Equation 10 suggests the possibility of decomposing a large structuring element by repeated dilation with a 3 x 3 Small Structuring Element (SSE) in analogy to the SGK decomposition of an impulse response. Figure 15 illustrates several such decompositions. A decomposition exists for structuring elements that are convex and contain no interior zeros, but unfortunately, no analytic decomposition procedure has yet been developed.

```
                                           1 1 1 1 1
        1 1 1        1 1 1                 1 1 1 1 1
        1 1 1   (+)  1 1 1     =           1 1 1 1 1
        1 1 1        1 1 1                 1 1 1 1 1
                                           1 1 1 1 1


                                           0 1 1 1 0
        1 1 1        0 1 0                 1 1 1 1 1
        1 1 1   (+)  1 1 1     =           1 1 1 1 1
        1 1 1        0 1 0                 1 1 1 1 1
                                           0 1 1 1 0


                                           0 0 1 0 0
        0 1 0        0 1 0                 0 1 1 1 0
        1 1 1   (+)  1 1 1     =           1 1 1 1 1
        0 1 0        0 1 0                 0 1 1 1 0
                                           0 0 1 0 0


                                           0 0 1 1 1
        0 1 1        0 1 1                 0 1 1 1 1
        1 1 1   (+)  1 1 1     =           1 1 1 1 1
        1 1 0        1 1 0                 1 1 1 1 0
                                           1 1 1 0 0
```

Figure 15   Examples of small structuring element decomposition.

8.7   SUMMARY

It has been shown that a wide variety of image processing and analysis algorithms can be decomposed into a series and parallel interconnection of primitive computational elements. This decomposition permits rapid computation in an image computer with a pipeline architecture.

8.8 REFERENCES

1. W.K. Pratt and P.F. Leonard, "Review of Machine Vision Arch—itectures", Proceedings SPIE Conference, Los Angeles, California, January 1987.

2. J.F. Abramatic and O.D. Faugeras, "Design of Two Dimensional FIR Filters from Small Generating Kernels," Proceedings IEEE Conference on Pattern Recognition and Image Processing, Chicago, Illinois, May 1978.

3. W.K. Pratt, J.F. Abramatic, and O.D. Faugeras, "Method and Apparatus for Improved Digital Image Processing", United States Patent 4,330,833, May 18, 1982.

4. S.U. Lee, "Design of SVD/SGK Convolution Filters for Image Processing," University of Southern California, Image Processing Report USCIPI No. 950, January 1980.

5. W.K. Pratt, "Intelligent Image Processing Display Terminal", Proceedings SPIE, Vol. 199, San Diego, California, August 1979, pp. 189–194.

6. W.K. Pratt, Digital Image Processing, Wiley Interscience, New York,1978.

7. R.H. Wallis, "An Approach for the Space Variant Restoration and Enhancement of Images", Proceedings of Symposium on Current Mathematical Problems in Image Science, Monterey, California, November, 1976.

8. K.I. Laws, "Textured Image Segmentation", University of Southern California, Image Processing Institute, Report USCIPI 940, January 1980.

9. S.R. Sternberg, "Biomedical Image Processing", IEEE Computer, January 1983, pp. 22–34.

10. R.M. Haralick, S.R. Sternberg and X. Zhuang, "Image Analysis Using Mathematical Morphology", IEEE Transactions on Pattern Analysis and Machine Intelligence, Vol. PAMI–9, No. 4, July 1987, pp. 532–550.

9

EDGE EXTRACTION AND LABELLING
FROM STRUCTURED LIGHT 3–D VISION DATA

H.S. YANG AND A.C. KAK
Robot Vision Lab
School of Electrical Engineering
Purdue University

9.0 ABSTRACT

In this paper we first discuss various structured light scanning strategies, some of these being particularly suitable for integration with robot manipulation for the acquisition of 3–D vision data. We have supplied formulas for converting structured light information into range maps for each scanning procedure. We then present a novel scheme for extracting edges from range maps. Our scheme consists of first converting a range map into a needle diagram and constructing from the needle diagram two synthetic gray level images; the gray levels in one are proportional to needle magnitudes, while in the other to orientations. From these synthetic gray level images, which are virtually noise–free, we can obtain object edges by applying any gray level edge detector. The extracted edges are then labeled *convex, concave, outer obscuring and interior obscuring*, such labels being useful for scene segmentation.

9.1 INTRODUCTION

Recently, range data has emerged as a necessary input for both reliable shape recognition of 3–D objects, and determination of their locations and orientations [4,8,13,14,16,21,22,26]. In the past, shape recognition schemes were based mostly on silhouette and/or line representation of objects obtained from photometric (reflectance) information; a silhouette is the boundary of the object against its background, and a line is usually caused by an edge formed at the intersection of two faces. It is often difficult to reliably extract silhouettes and lines from the reflectance data because of the problems caused by brightness–sensitivity variations across the sensor, electronic noise, contrast reduction and shadow effect; the last two, namely contrast reduction and shadow effect, are the most challenging to deal with. Contrast reduction, often caused by multiple reflections between surfaces if they are facing each other, leads to diminished brightness change across an edge; whereas the shadow effect generates false contrast. Under these circumstances, 2–D difference edge operators either fail to detect edges or produce spurious ones. The quality of results obtained can be improved by using high contrast backgrounds (black for bright objects and white for dark objects), an equivalent effect being obtained by locating objects on a light table. Although this does help with the outer boundaries, there still remains the annoying problem of extracting interior boundaries that correspond to the intersections of surfaces at different orientations, especially if those boundaries are critical to a meaningful analysis of the scene. There do exist sophisticated procedures which, at least

148

in theory, can retrieve missing edges by using dictionaries and other forms of knowledge bases.   It is usually most difficult to remove false edges without also damaging the true ones.

Many of the aforementioned difficulties with the extraction of external and internal boundaries disappear when range data is used, either by itself or in conjunction with reflectance data [6,7,16].   In this paper, we will only be concerned with extracting, for the end purpose of scene analysis, all the desired low level information from just the range data.   In this context, it was first pointed out by Nitzen et al. [16] that there are two types of edges in the range map of a scene.   The first type, called a jump boundary, corresponds to the boundary of an occluding object, which causes discontinuities in the measured range data.   Jump boundaries are easily detected by searching for discontinuities in the range data that exceed a certain threshold.   The second type of edge in range data is called a "roof edge" and corresponds to a junction of surfaces characterized by different orientations. Roof edges are harder to detect, since they require that one, directly or indirectly, measure the changes in the surface characterizing parameters.

It is the extraction of these latter type of range data edges, namely the roof edges, that this paper makes a main contribution towards.   Inokuchi and Nevatia [10] detected edges in the range data by convolving the range map with ideal edge masks in six directions that were 30^0 apart.   Since the edge masks used were suitable only for detecting jump boundaries, the performance of their operators for detecting roof edges was less than reliable. Gil, Mitiche and Aggarwal [7] have extracted edges from range data by roughly estimating at every point in the range map the curvature of the local surface and thresholding the result.   The point of their exercise was to use the extracted range edges for providing supporting evidence for edges obtained from the reflectance data.   Sugihara [22] has used special operators to first locate edges in the structured–light range data and then used a dictionary to predict the position, orientation, and physical characteristics of missing edges. In Sugihara's approach, the need for prediction arises since his operators are capable of generating complicated results around vertices.   Without prediction, some edges around a vertex can be missed altogether.   Moreover, since the curvature of a stripe along an edge varies according to the attitude of the object and the angle of the projected beam, with Sugihara's operators edges extracted even from non–vertex areas may not always be reliable, making necessary a dictionary–based approach.

In the approach presented here, we conceptualize edges as intersections of planar or curved surface regions.   Therefore, the location of an edge in our algorithms is characterized by what is known about the orientation of adjacent surface patches.   After the surface orientations are computed from a range map, the first step in our procedure is the creation of a synthetic gray–level image from the surface orientation data; and then application of any typical 2–D edge operator to this synthetic image yields the edges present in the scene. *In actuality, since the orientation of a surface in 3–D space has two degrees of freedom and since the gray level gives us only one parameter for representation, for each scene we construct two synthetic gray level images that correspond to the two degrees of freedom; an edge in the scene must show up as a gray level discontinuity in one of these two synthetic images.* In general, the four factors — the light source direction, viewing direction, surface reflectance property and the surface orientation — determine the brightness of a point on a surface. Assuming that the first three factors can be held constant, the perceived brightness is only a function of local surface orientation, which would cause the perceived contrast at an edge to be only a function of the difference in the orientations of the surfaces which meet at that edge. Therefore, it would seem that our approach captures this aspect of human perception, since an edge is declared to be present at a given location if there exists a computed difference between the orientations of the adjoining surfaces at that point.

In what follows, in Section 9.2 we will first discuss different aspects of the acquisition of range data by structured light. We feel that an understanding of the underlying data collection mechanisms is important for the design of efficient low level operators; this statement is particularly true for structured light, since in such systems one can directly use what is called the "pixel offset information," instead of absolute range values for input to low level operators. As was pointed out by Oshima and Shirai [17], the pixel offset data is obtained very simply, without any computations or image digitization, by hooking a peak detector to the back of a camera. Since some people are disheartened by the fact that in a structured light system the data collection time may be too long for certain applications, as one has to scan the scene one stripe at a time, we have also discussed in this section some strategies for alleviating that problem.

We will then begin Section 9.3 with a discussion of the computation of surface orientation information from the range data, paralleling what has been done before by Shirai et al [17,21]. This will be followed by the presentation of how we generate synthetic gray–level images from the surface orientation information. We will conclude this section with some comments on the data bandwidth required and the computational complexity involved.

Finally, in Section 9.4 we will present a discussion on the extraction of edges from the synthetic gray level images. There we talk about various edge types and the construction of 8–connected edges that are thinned and labelled.

9.2   STRUCTURED LIGHT FOR THE ACQUISITION OF RANGE DATA

It is relatively easy to design robust and fast algorithms for 3–D vision using structured light data; in addition, systems for generating such data tend to be rather inexpensive. In Section 9.2.1, we will first discuss a single slit projection system for acquiring structured light data; and then in Section 9.2.2 show how the data collection time can be reduced by projecting parallel grids of encoded light onto the scene.

9.2.1   SINGLE SLIT PROJECTION SYSTEM

Agin and Binford [1] used a laser based structured light system capable of projecting light stripes at different orientations at the scene; the resulting data was then transformed into generalized–cylinder models for the objects present in the scene. Oshima and Shirai [17] have reported a system that uses a rotating mirror to scan the scene with a thin stripe of light; the range maps so obtained are fitted with surfaces consisting of planar patches, each patch defined over a small set of points. Based on the smoothness of surface normal variation, the planar patches are then aggregated into larger plane and curved surfaces.

In this section, we will present three different approaches to structured light scanning for the purpose of building up a range map; two of these have been implemented in our lab — one of the two being particularly suited for integrating structured light sensing with robot manipulation. These three approaches will be referred to as

1.   Fan Scanning
2.   Type–1 Linear Scanning
          (appears to be ideal for integrating with robot manipulation)
3.   Type–2 Linear Scanning

Fan scanning with a rotating mirror and Type–1 linear scanning with a Cincinnati–Milacron T^3–726 robot arm are currently in operation in our lab.

A Fan Scan is similar to what has been reported by Oshima and Shirai. As shown in Fig. 1a, a projector is used to generate a stripe of light, which is then reflected off a mirror on to the scene, the mirror being under computer control which makes it possible to scan the scene in a step by step fashion, or, for that matter in any other desired manner within the limitations of the fan–scan configuration. For each stripe, the position of the pixel with maximum brightness in each scan line of the camera is determined by a custom designed peak detector attached to the back of the camera. Fig. 1b depicts a Type–1 Linear Scan, which is most easily accomplished with the help of a manipulator with its gripper holding a sensor consisting of a projector and a camera. In the simplest form of Type–1 scanning, the sensor moves in a straight line and records the illuminated stripes at equal intervals

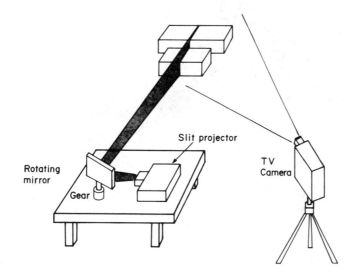

Figure 1(a)   A single slit projection system for fan scanning

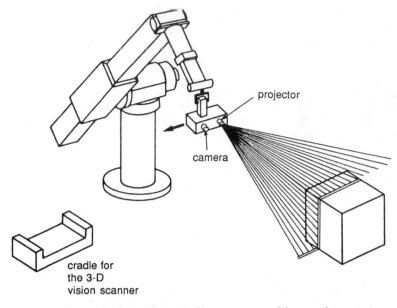

Figure 1(b)   Type–1 linear scan with a robot.

along the direction of motion.   Variations on this approach are possible, such as adaptive scanning (fewer stripes in regions where there are no objects), 'looks' from different angles, etc.   Fig. 1c depicts a Type–2 Linear Scan, which differs from the Type–1 Linear Scan in only one respect: a fixed

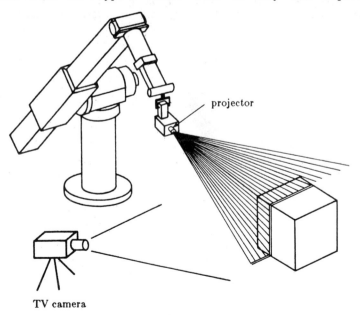

Figure 1(c)   Type–2 linear scan with a robot.

camera is now used to record the illuminated stripes that are projected from a moving source in the gripper of a manipulator.

Note that in all scanning strategies, other factors being constant, the width of a projected stripe in the scene depends upon the width of the slit in the projector slide. If the slit is too wide, the resulting wide stripes may break up and lead to multiple peaks of brightness along each raster line in the camera image; on the other hand, if the slit is too narrow, the amount of light projected onto the scene may not be enough to cast a perceptible image on the camera. Optimum slit widths are a trade off between these two considerations and are obtained by trial and error. Shown in Fig. 2a is an example of an illuminated stripe in a 3–D scene consisting of a toy robot placed on top of a book. Fig. 2b shows the light stripe image as captured

Figure 2(a)   A scene consisting of a toy robot and a book.

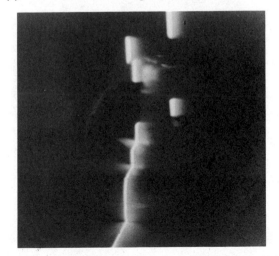

Figure 2(b)   A single light stripe image taken under dark conditions.

by the frame grabber.   Figs. 3a and 3b depict, respectively, the thresholded and the thinned stripes.

The stripe images as recorded on the camera can be translated easily into what is called the pixel offset data.   In Fig. 4 we have shown what is meant by pixel offset data.   In that figure, with pixel P in the source–viewpoint frame, we associate the offset $d(i,j)$ as obtained from the location of the corresponding illuminated pixel in the camera image.   In the rest of this section, we will illustrate how the pixel offset data is translated into range maps for each of the scanning strategies.

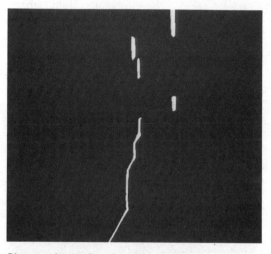

Figure 3(a)   Shown here is the binary image obtained by using a threshold of 170.

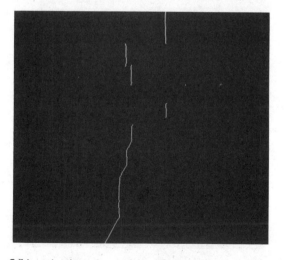

Figure 3(b)   A tinned version of the binary stripe image.

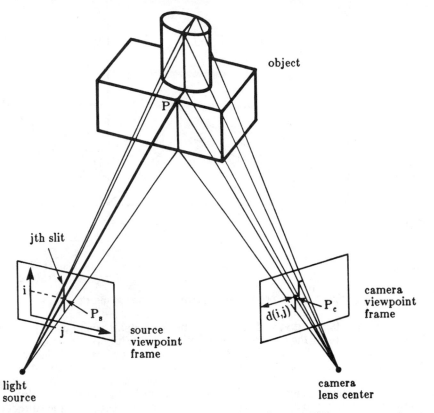

Figure 4   Shown are the source viewpoint frame (light source plane) and the camera viewpoint frame (camera image plane). Pixel offset d(i,j) is the horizontal distance in the camera image corresponding to point P on the j–th stripe projected by the source. The quantity d(i,j) is measured from the left hand edge of the i–th scan line of the camera image.

9.2.1.1   FAN SCANNING

Fig. 5 illustrates how the coordinates (x,y,z) of an object point P are computed from the following triangulation formulas:

$$z_o = \frac{d}{\cot(\theta_m) + \cot(\theta_c)}$$

$$x_o = z_o \cot(\theta_m)$$

$$y_o = \sqrt{(x_o - d)^2 + z_o^2}\ \tan(\theta_d)$$

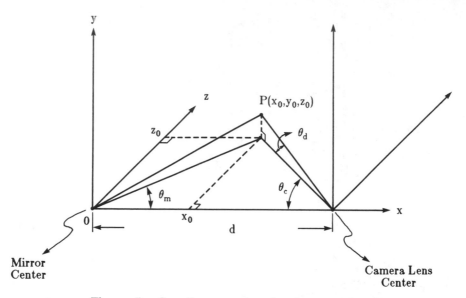

Figure 5 Coordinate system for fan scanning.

Note that θ_m is given directly by the rotational position of the mirror, which is known since the angle of the mirror in Fig. 1a is under computer control. The angles θ_c and θ_d are determined from a knowledge of the camera optic axis and the location of P (Fig. 5) in the camera image. The direction of the camera optic axis may be obtained by going through a calibration procedure, such as the one outlined in [13], or by actual measurement, assuming that there is no great disparity between the direction of the optic axis and the longitudinal axis of the camera box, as visible from the outside.

We will now describe an alternative method for computing the object point coordinates that does not require the use of transcendental functions. As described in [2], the transformation C (camera matrix) from world point P(x,y,z) to image point (U,V) can be written as

$$
\begin{bmatrix} u \\ v \\ w \end{bmatrix} = C \begin{bmatrix} x \\ y \\ z \\ 1 \end{bmatrix}
\tag{1}
$$

where

$$
U = \frac{u}{w} \quad \text{and} \quad V = \frac{v}{w}
\tag{2}
$$

and C is a 3×4 matrix. In practice, the matrix C can be found by showing to the camera a few object points whose world coordinates are known.

With C known and with the image point (U,V) given, the direction of a line of sight, (λ,μ,ν), to the corresponding object point can be obtained by the following inverse perspective transformation:

$$\lambda = b_1 c_2 - b_2 c_1 \qquad (3)$$
$$\mu = c_1 a_2 - c_2 a_1$$
$$\nu = a_1 b_2 - a_2 b_1$$

where

$$a_1 = UC_{31} - C_{11}, \; a_2 = VC_{31} - C_{21}$$
$$b_1 = UC_{32} - C_{12}, \; b_2 = VC_{32} - C_{22}$$
$$c_1 = UC_{33} - C_{13}, \; c_2 = VC_{33} - C_{23}$$

and C_{ij} is (i,j) element of the camera matrix C.

As illustrated in Fig. 6, if the coordinates of P_c, the camera lens center, are known, then the coordinates of the world point P are determined by the intersection of the line

$$\frac{x-x_c}{\lambda} = \frac{y-y_c}{\mu} = \frac{z-z_c}{\nu}$$

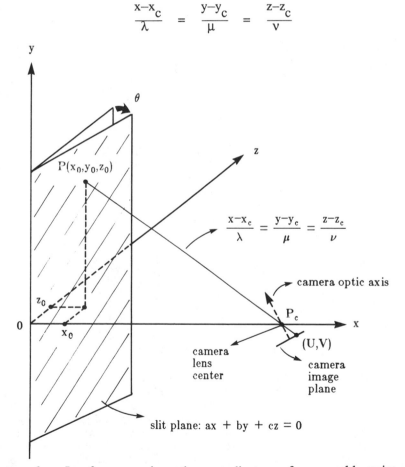

Figure 6   In fan scanning the coordinates of a world point at (x_0, y_0, z_0) are given by the intersection of the light stripe plane and the line of sight from the camera.

with the plane $ax + by + cz = 0$, where (x_c, y_c, z_c) are the coordinates of the camera lens center. In the equation of the plane, the surface normal of a slit–plane rotated about the y axis by θ from the plane is represented by (a,b,c). These direction cosines with respect to the work–platform plane, whose normal is $(1,0,0)$, are given by

$$\begin{bmatrix} a \\ b \\ c \end{bmatrix} = \begin{bmatrix} \cos\theta & 0 & \sin\theta \\ 0 & 1 & 0 \\ -\sin\theta & 0 & \cos\theta \end{bmatrix} \begin{bmatrix} 1 \\ 0 \\ 0 \end{bmatrix}$$

or, equivalently, by

$$\begin{aligned} a &= \cos\theta \\ b &= 0 \\ c &= -\sin\theta \end{aligned}$$

Therefore, one can write the following expressions for the coordinates of the object point

$$\begin{aligned} x_o &= x_c + t\lambda \\ y_o &= y_c + t\mu \\ z_o &= z_c + t\nu \end{aligned}$$

where

$$t = -\frac{x_c\cos\theta - z_c\sin\theta}{\lambda\cos\theta - \nu\sin\theta}$$

Without knowing the coordinates, P_c, of the camera lens center, it is still possible to compute the coordinates, P, of the object point. Since P is the intersection point of a slit plane and the two planes whose intersection constitutes the line of sight, the coordinates of P will be given by

$$\begin{bmatrix} x_o \\ y_o \\ z_o \end{bmatrix} = \begin{bmatrix} a & b & c \\ a_1 & b_1 & c_1 \\ a_2 & b_2 & c_2 \end{bmatrix}^{-1} \begin{bmatrix} 0 \\ d_1 \\ d_2 \end{bmatrix}$$

where

$$\begin{aligned} d_1 &= UC_{34} - C_{14} \\ d_2 &= VC_{34} - C_{24} \end{aligned} .$$

9.2.1.2 TYPE–1 LINEAR SCANNING

As mentioned earlier, Type–1 Linear Scan, depicted in Fig. 1b, is ideal for integrating the range mapping process with manipulation; also, it allows the data collection to be tailored to the goals of the manipulation task. Fig. 7 illustrates the parameters involved in the calculation of object point coordinates, (x_0, y_0, z_0), from the image points recorded on the camera. Note that the slit plane now keeps a fixed orientation as it is translated to the right by a distance x_0 together with the camera. We will assume that the current sensor (combination of the projector and the camera) position corresponds to the shaded slit plane and the displaced camera lens center (at $x_c + x_0$) in

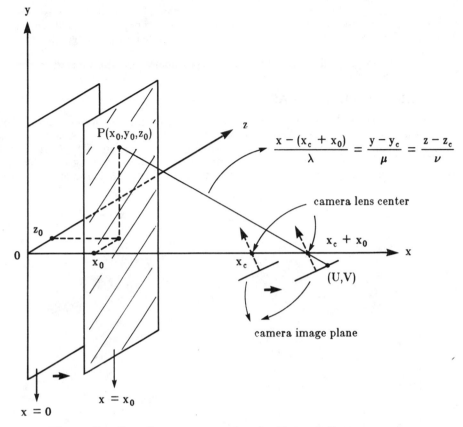

Figure 7   Coordinate computation in Type–1 linear scanning.

Fig. 7.   The object point coordinates are given by the intersection of the slit plane with the line of sight:

$$x_0 = k\Delta x$$

$$y_0 = y_c - \frac{\mu x_c}{\lambda}$$

$$z_0 = z_c - \frac{v x_c}{\lambda}$$

or, equivalently,

$$y_0 = \frac{c_1 d_2 - c_2 d_1}{b_1 c_2 - b_2 c_1}$$

$$z_0 = \frac{b_2 d_1 - b_1 d_2}{b_1 c_2 - c_1 b_2}$$

where k is the index of the light stripe and Δx is the distance between successive locations of the sensor along the x–axis.

One unique advantage of this type of scanning is that for most scenes, it probably suffers the least from the missing parts problem. Furthermore, it is now possible to devise a table look–up for translating image coordinates directly into object coordinates. These tables are indexed by the x–coordinate of each projection, and then for the (U,V) coordinates of illuminated points in the camera image they directly yield the object point coordinates.

9.2.1.3 TYPE–2 LINEAR SCANNING

In the Type–2 linear scanning, a slit projector is moved along the x–axis while the camera is kept stationary at one location. Assuming that calibration parameters are known for the camera, as shown in Fig. 8, from this type of scanning the coordinates of the object point are given by the intersection of the slit plane with the line of sight from the camera:

$$x_o = k\Delta x$$
$$y_o = y_c + \frac{\mu(x_o - x_c)}{\lambda}$$
$$z_o = z_c + \frac{\nu(x_o - x_c)}{\lambda}$$

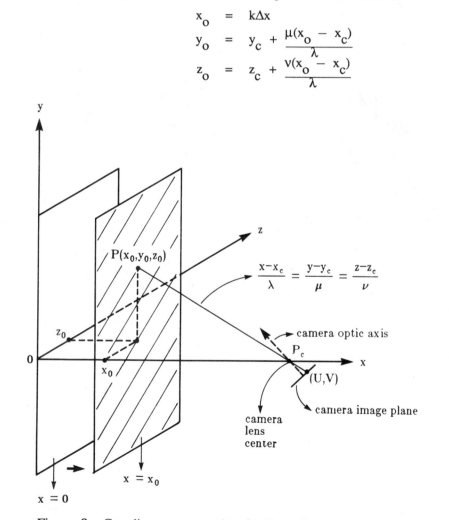

Figure 8   Coordinate computation in Type–2 linear scanning.

where k is the index of the projection and Δx is the shift of the slit plane along the x axis per projection. Note that this type of scanning is amenable to fast computation since no trigonometric formulas are involved.

If the location of camera, as represented by P_c, is not known precisely, one can use the following formulas:

$$x_o = k \; \Delta x$$

$$y_o = \frac{(a_2c_1 - a_1c_2)x_o + (c_1d_2 - c_2d_1)}{b_1c_2 - b_2c_1}$$

$$z_o = \frac{(a_1b_2 - a_2b_1)x_o + (b_2d_1 - b_1d_2)}{b_1c_2 - c_1b_2}$$

Before concluding this section, we will now make a few remarks about factors that affect the accuracy of derived range maps. In Fig. 9a, is shown a photometric image of a scene consisting of two overlapping boxes, and Fig. 9b shows the light stripe image as seen by the camera. The light stripe distances from the left edge of the display in Fig. 9b are the offsets $d(i,j)$. The accuracy of a range map is dependent on the baseline distance between the mirror center and the camera lens center; the larger this distance the more accurate the range measurements. However, as the baseline distance increases, there is a greater occurrence of the occlusion problem; meaning there come into existence regions that are visible from the mirror (light source) viewpoint, yet not visible from the camera viewpoint. This occlusion problem, which is shared by all ranging schemes based on triangulation, causes us to not know the range data for portions of a scene. The line segment AC in Fig. 10, for instance, is hidden from the camera by the front box, so the range data associated with this line segment cannot be obtained. On the other hand, the line segment BD is not illuminated by the source; the range data for these points as well will not be obtained.

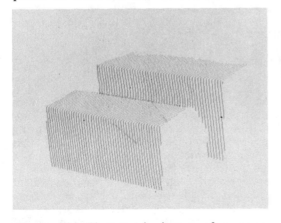

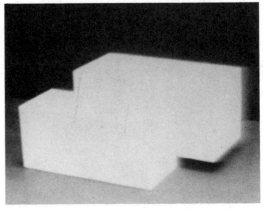

Figure 9(a) Photometric image of a scene consisting of two overlapping boxes.

Figure 9(b) Light stripe image of (a).

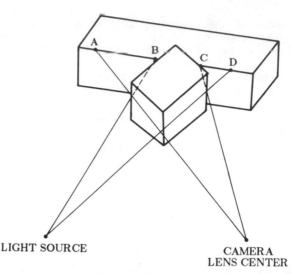

LIGHT SOURCE

CAMERA
LENS CENTER

Figure 10 . The missing part problem in triangulation based methods for range mapping is depicted here. It is not possible to project a light stripe on BD, and AC is hidden from the camera.

A disadvantage of the single slit projection system is the time it takes to scan a scene for acquiring its range map. It has been suggested [18] that perhaps a scene could be scanned by more than one slit at a time which would then reduce the data collection time by a factor equal to the number of slits used. The main difficulty with this approach is that the sequence in which stripes appear in a given region of the scene may not be the same as the sequence in which they are projected; this can generate ambiguities in stripe identification.

In the next section, we will show that better procedures do exist for getting around the scan time difficulties associated with a single slit projection system; these procedures do not suffer from the above mentioned ambiguities in stripe identification, and consist of projecting coded parallel grids on the scene. In the context of range mapping, such schemes were first advanced by Minou [15].

9.2.2   CODED PARALLEL GRID PROJECTION SYSTEMS

In a coded parallel grid projection system, we illuminate a scene with a number of coded grid patterns, like those shown in Fig. 11, as opposed to scanning the scene with a single slit. Although the scene is now being illuminated with patterns, from the resulting photometric images we still want to be able to extract information identical to what would be obtained with a single slit scan. The illumination (or projection) patterns are constructed by using the following procedure: We represent each stripe in an equivalent single slit scan by a binary coded word, the bit planes built out of these code words are the illumination patterns; the implication being that for N stripes we would need to project $\log_2 N$ patterns on the scene (assuming N is a power of 2). If N is not a power of 2, the number of projections required is given by int($\log_2 N$)+1, where by int(X) is meant the integer part of the real number denoted by X.

From the photometric data for parallel grid projections, the individual stripes corresponding to an equivalent single slit scan are obtained by a simple decoding algorithm which consists of examining for a point in the scene the on–off sequence of illuminations obtained for all the projected grid patterns, and then placing there a stripe corresponding to the resulting binary code word.

Fig. 11 illustrates 4 parallel grid patterns that yield 16 stripes for range mapping the scene. Fig. 12 illustrates the fact that each of the 16 stripes has been decoded from the four photometric images of a scene consisting of a cylinder resting on a horizontal surface. For the purpose of display only, the sixteen stripes in Fig. 12 are represented by different gray levels, indicating that each was separated from the rest and labeled correctly.

Notwithstanding the reduction in the data collection time, the method of projecting parallel grids does suffer from a couple of rather serious disadvantages. The first one has to do with the registration of different grid patterns — a necessary prerequisite to the decoding process. For example, for the case of the four projection patterns depicted in Fig. 11, any mechanical jitter in the projection system would cause the patterns to become misaligned and lead to errors in stripe decoding. A certain degree of insensitivity to misalignment can be built in by incorporating delimiting spaces between opaque and transparent regions of the patterns and hypothesizing the existence of similar delimiting spaces between the individual stripes of an equivalent single slit scan. Other solutions to overcome this problem include the use of time coded parallel grid projections employing a facsimile [15] and a Gray–coded pattern projector using an electro–optical device for the structured light source [12].

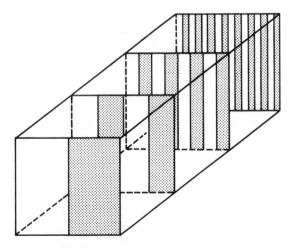

Figure 11 Four binary coded grid patterns required for labelling 16 stripes uniquely. No delimiting spaces exist between stripes.

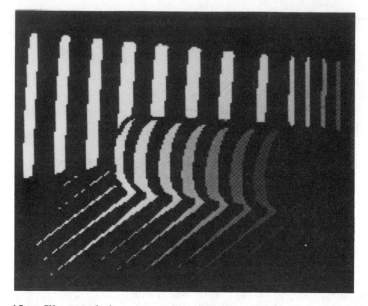

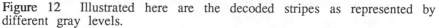

Figure 12 Illustrated here are the decoded stripes as represented by different gray levels.

The second, and more serious, difficulty with the method of projecting parallel grids has to do with the selection of suitable thresholds for stripe detection. With a single slit scan, all one has to do for detection is to locate the maximum brightness in a video raster line (although, the problem really isn't as simple as that due to the fact that occasionally the stripes wrap around objects leading to more than one stripe in a single video line even though only a single slit is used for illumination, but situations like that are rare). When full frame patterns are projected on the scene, the detection of pattern components becomes a challenging problem in image segmentation; problems of that kind have defied easy solutions in the past. In the rest of this section, we will demonstrate more clearly the difficulty that arises when only simple procedures are used for thresholding, and then present a method that has yielded good results in our laboratory.

Stripe detection by thresholding is made complicated by the uneven reflectance of object surfaces in the scene and the uneven focusing of the projection device. Fig. 13a shows a scene consisting of a gray colored book on the table and Fig. 13b is light stripe image of this scene taken under the normal room lighting conditions. Fig. 14a is a gray level histogram of Fig. 13b; since this histogram does not have a distinguished valley, it is impossible to choose a global threshold to detect stripes. The best thresholded stripe image is obtained by setting the threshold at 200 and is shown in Fig. 14b. Fig. 14c is the histogram of the light stripe image of the same scene taken under conditions when the room light was turned off. The best threshold is 190 and the thresholded image is shown in Fig. 14d. An examination of these histograms and the detected stripes indicates that the gray level distribution of a light stripe image is unpredictable and depends upon room lighting conditions.

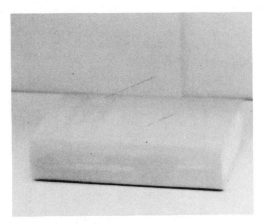

Figure 13(a)   A gray colored book on a table.

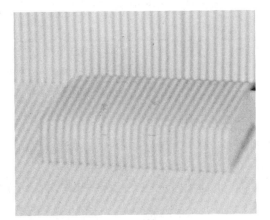

Figure 13(b) Light stripe image of the scene taken under normal room lighting conditions.

It is possible to obtain a superior delineation of stripes if instead of a global threshold derived from a histogram we use only local information, one example of which would be to use a local threshold.   In their implementation of local thresholding, Inokuchi et al. [12] compared a light stripe image with a photometric image of the scene (with the projector turned off) taken under normal lighting conditions, and detected the stripes by checking whether the brightness of a point in the light stripe image was greater than that of the corresponding point in the photometric image.   Despite the local nature of thresholding, the results are not entirely satisfactory; as is evidenced by Fig. 15 obtained by comparing the light stripe image in Fig. 13b with the reflectance image of Fig. 13a, and retaining only those pixels whose gray levels in the former are greater than those in the latter.   It is not difficult to figure out why this method may not perform well: the reason being that when the projector is turned on, the shadow effects in the scene are in general different

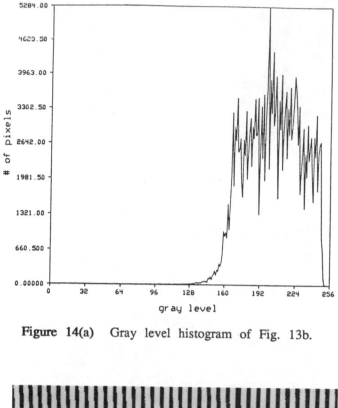

Figure 14(a)   Gray level histogram of Fig. 13b.

Figure 14(b)   The best looking binary stripe image is obtained by using a threshold of 200.

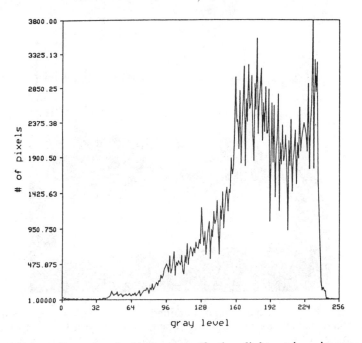

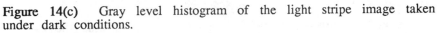

Figure 14(c) Gray level histogram of the light stripe image taken under dark conditions.

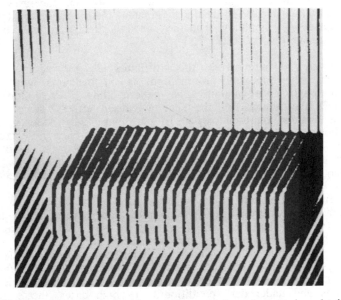

Figure 14(d) The best looking binary stripe image is obtained by using a threshold of 190.

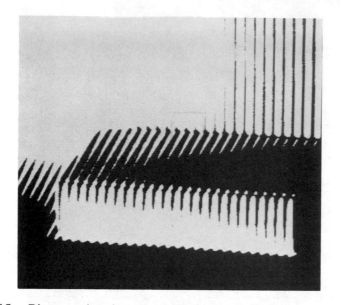

Figure 15   Binary stripe image obtained by comparing the photometric image taken under normal lighting conditions (Fig. 13a) with the light stripe image of Fig. 13b.

from those under normal room lighting conditions. Comparing the two images in Fig. 13, one can notice in the top left region that the gray levels there associated with the black stripes are greater than the gray levels in the photometric image — implying that the shadow effects are not the same with the projector turned on and off.

One way to circumvent this difficulty is to record the reference reflectance data when not only room light is on but also with the projector turned on, although not containing the pattern slide. Projection of a blank slide guarantees the removal of the shadow–caused effects, since such shadows now more accurately represent what is out there when the grid pattern is projected. For a more accurate recording of the reference reflectance data, we suppose one should use a blank slide with transmittance equal to the average transmittance of the pattern slide; this we have not done. Fig. 16a shows a reference image thus recorded and in Fig. 16b are shown the detected stripes.

To further illustrate that superior results are obtained when local thresholds are derived from the photometric image recorded with the light projector on (with a blank transparent slide), we will show detected stripes for an object with a textured surface that is rather poorly reflecting. The object in Fig. 17a has white figures and characters on a dark blue cover and Fig. 17b is its light stripe image. Figs. 18a, b, c and d show the gray level histograms and the best thresholded stripe images obtained under the room light conditions and under dark conditions. In both cases, the extracted stripes are of poor quality. On the other hand, Fig. 19 is obtained by comparing the reference image of Fig. 17a and the stripe image of Fig. 17b; note that the reference image was recorded with the projector turned on and containing a blank slide.

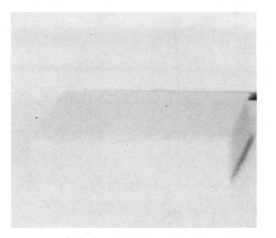

Figure 16(a) Reference image taken when the room lighting is on; also on is the projector with a blank transparent slide in it.

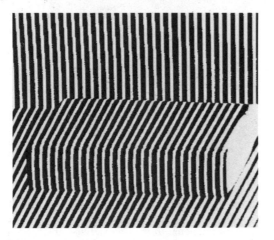

Figure 16(b)   Binary stripe image obtained by comparing the reference image of Fig. 16a with the light stripe image of Fig. 13b.

Now that we have discussed various methods for constructing range maps via the use of structured light, in the next section we will address the issue of automatic interpretation of these range maps. For all the experimental results shown there, the range data was obtained with single slit scanning.

9.3   SURFACE SHADING BASED ON NEEDLE ORIENTATION AND MAGNITUDE INFORMATION

9.3.1   COMPUTER REPRESENTATION OF RANGE MAPS

In deciding how the range maps as obtained by structured light might be represented in computer memory, one has to bear in mind that there are two viewpoints to contend with.   As shown in Fig. 4, there is the source

Figure 17(a)   Reference image of a book that has white figures and printing on a dark blue cover.

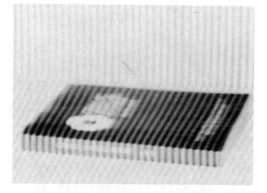

Figure 17(b)   Light stripe image of the scene.

viewpoint corresponding to an observer stationed at the light source, and then there is the camera viewpoint.   The occlusions, as perceived from the source viewpoint would, in general, manifest themselves differently in what is seen from the camera viewpoint.   (This never creates a problem with how the computer processes the data, but only in how we might picture the workings of a computer program, since we are more at ease with what is perceived from the camera viewpoint, whereas the data that the computer works on corresponds to the source viewpoint).

From the geometry shown in Fig. 4, it is readily apparent that from the source viewpoint the position of each projected stripe of light can be represented by a single index, which in fan–scanning corresponds to the rotational angle of the stepping motor used for turning the mirror.   The raw data therefore consists of a matrix whose columns correspond to the individual stripes projected onto the scene, and rows to raster scans.   At each location of this matrix is stored a set of 4 numbers: the pixel offset data and the x, y, and z coordinates of the corresponding illuminated point in the scene.   If this information were to be displayed in a straightforward manner, meaning as an array with the brightness of each element proportional to the corresponding

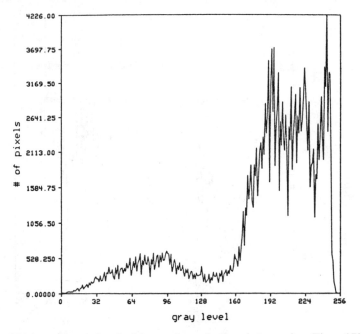

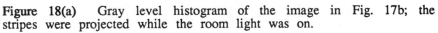

Figure 18(a)   Gray level histogram of the image in Fig. 17b; the stripes were projected while the room light was on.

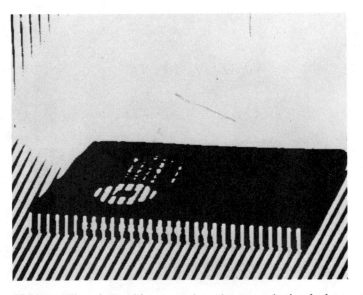

Figure 18(b)   The best binary stripe image obtained by using a threshold of 170.

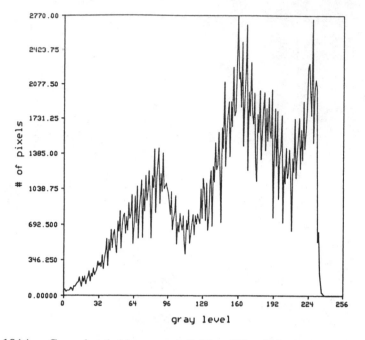

Figure 18(c) Gray level histogram of Fig. 17b. Stripes were projected with the room light off.

Figure 18(d) The best thresholded stripe image obtained by using a threshold of 160.

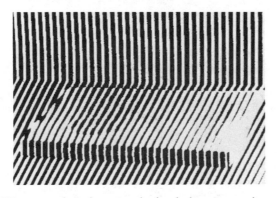

Figure 19   Binary stripe image obtained by comparing the reference image of Fig. 17a with the light stripe image of Fig. 17b.

$d(i,j)$, the result would be unintelligible.   The matrix is best displayed from the camera viewpoint with $d(i,j)$ being what they are, pixel offsets.   This is how the display in, for example, Fig. 9b was made, even though each stripe shown there is a single column of the matrix stored in the computer.

Although there is ample precedence for representing range maps as raster–scan x stripe matrices in the manner discussed above [17], it is also possible to store the range information as a matrix that directly corresponds to what is perceived from the camera viewpoint.   In most cases, the number of stripes collected for a scene is far less than what would correspond to the resolution of the camera; in most of our experiments, we have only used 100 stripes.   This implies that along a raster line in the camera frame the information $d(i,j)$ is not available at every pixel.   As was done by Tomita and Kanade [23], this can be rectified by interpolation, but the end result is a large sized, say 512×512, range map for the case where one might only have 100 stripes illuminating the scene.   The intuitive solution of sampling the camera frame over a smaller set of points, say 100×100, is really not applicable, since then the error generated when the illuminated points do not fall on the sampling grid of camera frame would be unacceptable.   Note that in the stripe–by–raster scan representation we only have to store a 512×100 matrix, assuming a 512×512 discrete representation for the camera image.

9.3.2   FITTING PLANAR PATCHES BY THE LEAST SQUARES METHOD

The next step is to fit planar patches to adjoining pixels in the scene. Following Shirai's procedure [17], we cluster pixels into overlapping 8×8 regions as shown in Fig. 20.   For a structured light scan consisting of 100 stripes, the range data consists of a 480×100 matrix conforming to the format described in the preceding section.   Using 8×8 overlapping planar patches gives us a total of 120×50 patches for the scene.   Note that on account of the fact that the physical distance between the stripes illuminating a surface depends upon the orientation of that surface, the 8×8 patches will not, in general, correspond to square regions on object surfaces.   Fig. 21 is a depiction of the arrangement of surface patches on a hypothetical scene.

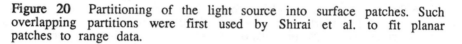

Figure 20 Partitioning of the light source into surface patches. Such overlapping partitions were first used by Shirai et al. to fit planar patches to range data.

Not all points in a 8×8 cluster may be used for planar patch fitting, since in some cases the cluster may be straddling an edge where two different surfaces join.   Also, in some other cases, an 8×8 region can overlap two different surfaces that are separated by a depth discontinuity.   Since in both these situations, fitting a planar patch to the entire 8×8 cluster would be a wrong thing to do, we use only those points that can be considered to be on the same plane surface.   The decision to discard points is made on the basis of the deviation of the offset values from that of the center point of the patches.   Since the frame grabber digitizes the camera image into a 480×512 matrix, if we use 100 stripes to fill up the entire field of view, the maximum offset difference between the pixels on successive light stripes is five. Therefore, unless surface discontinuities and things of that sort are encountered, the maximum offset difference between the center of an 8×8 region and any of the peripheral points is around 20.   For this reason, we use a rejection threshold of 20.

We will now discuss the steps involved in fitting a planar patch to an M x M set of pixels.   In order to do so, we will assume the coordinate system of Fig. 5.   We will further assume that in this coordinate system a surface can be represented by the following form

$$z \;=\; f(x,y) \tag{4}$$

For a flat surface perpendicular to the z–axis, the surface normal is simply $(0,0,-1)$.   For any other surface the surface normal is given by

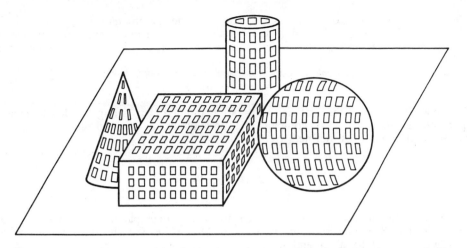

Figure 21   An example of local surface patch approximation to 3–D objects in a scene.

$$\left[\frac{\partial f}{\partial x} \ , \ \frac{\partial f}{\partial y} \ , \ -1 \right] \tag{5}$$

If the parameters p and q are defined as

$$p = \frac{\partial f}{\partial x} \ , \ q = \frac{\partial f}{\partial y} \tag{6}$$

then the surface normal can be written as

$$(p, \ q, \ -1) \quad . \tag{7}$$

It is more convenient to use a surface normal of unit length, which is given by

$$(p_n, \ q_n, \ r_n) \tag{8}$$

where

$$p_n = \frac{p}{\sqrt{1+p^2+q^2}} \tag{9}$$

$$q_n = \frac{q}{\sqrt{1+p^2+q^2}}$$

$$r_n = \frac{-1}{\sqrt{1+p^2+q^2}}$$

We will now show how one might calculate the parameters (p_n, q_n) for a set of M x M pixels. For this computation, we assume that the surface is locally plane over the M x M set; this assumption implies the following form for Eq. (4):

$$z = ax + by + c \tag{10}$$

The coefficients of this plane can be computed by using a least squares approach as follows. To compute the parameters, a, b, and c, we minimize the squared error J as defined by

$$J = \sum_{i=1}^{n} (z_i - ax_i - by_i - c)^2 \tag{11}$$

where n is the number of points used for the planar patch fit. Since not all points in the 8×8 clusters are used, $n \leq M^2$. We set the partial derivatives of J with respect to a, b and c to zero:

$$\frac{\partial J}{\partial a} = \sum_{i=1}^{n} (z_i - ax_i - by_i - c)x_i = 0$$

$$\frac{\partial J}{\partial b} = \sum_{i=1}^{n} (z_i - ax_i - by_i - c)y_i = 0$$

$$\frac{\partial J}{\partial c} = \sum_{i=1}^{n} (z_i - ax_i - by_i - c)z_i = 0$$

We thus obtain the following three equations for the three unknowns a,b and c:

$$a \sum_{i=1}^{n} x_i^2 + b \sum_{i=1}^{n} x_i y_i + c \sum_{i=1}^{n} x_i = \sum_{i=1}^{n} x_i z_i$$

$$a \sum_{i=1}^{n} x_i y_i + b \sum_{i=1}^{n} y_i^2 + c \sum_{i=1}^{n} y_i = \sum_{i=1}^{n} y_i z_i$$

$$a \sum_{i=1}^{n} x_i + b \sum_{i=1}^{n} y_i + cn = \sum_{i=1}^{n} z_i$$

From Eq.(6), p and q are identical to the coefficients a and b in Eq.(10). The above three equations therefore directly yield p and q; hence also, p_n and q_n.

9.3.3 NEEDLE ORIENTATION AND MAGNITUDE DIAGRAMS

For the purpose of display, the computed unit vectors for surface normals are projected on to the xy–plane (Fig. 5). It is easily shown that the projection of a vector (p_n, q_n, r_n) on the xy–plane is simply (p_n, q_n). The set of (p_n, q_n), as computed for all the patches, is called the needle diagram for the object, which is usually depicted as its orientation and magnitude.

In the needle orientation diagram, for each patch we compute the direction of the needle using $\tan^{-1}(q_n/p_n)$ and then show a unit vector in that direction.

To construct a discrete needle orientation diagram, we quantize 360^0 into 32 bins, from 0 to 31, with 0 assigned to the needle which is perpendicular to the image plane. The quantized values are termed the needle orientation code.

In Fig. 22b is shown the needle–orientation diagram computed from the light stripe image of Fig. 22a. Each needle in the diagram is a vector as defined by the pair (p_n, q_n). Similarly, Fig. 23b is the needle orientation diagram for the highly occluded scene consisting of five boxes (Fig. 23a).

Although needle orientation values can be uniformly quantized, the same cannot be said of needle magnitudes. As shown in Fig. 24a, if needle magnitudes are uniformly quantized from 0 to 1, different bins in the (p_n, q_n) space span significantly different areas on the Gaussian sphere; the bins farther away from the origin span Gaussian–sphere areas are larger than those closer to the origin. This difficulty can be reduced by quantizing the magnitudes sinusoidally (Fig. 24b). This means that we use the function $\sin\theta$ to set bin boundaries for magnitude values with θ — latitude on the Gaussian sphere — uniformly quantized between 0 and 90^0 (Fig. 24c). We have obtained better results with sinusoidal quantization, as opposed to uniform quantization for magnitude values.

For characterizing surface shape, it is possible to construct equi–orientation and equi–magnitude contours by connecting patches of equal orientation and magnitude, respectively. It is possible to classify surfaces as planar, spherical, cylindrical or conical by using these contours. Fig. 25 illustrates equi–orientation and equi–magnitude contours for three different

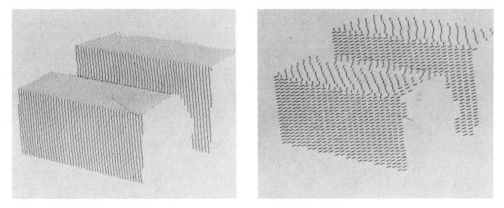

Figure 22(a)   Light stripe image for the scene in Fig. 9a.

Figure 22(b)   The needle orientation diagram obtained from the range map.

Figure 23(a)   A scene consisting of five highly overlapping boxes.

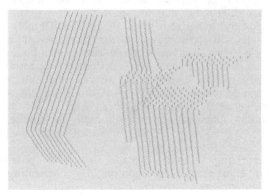

Figure 23(b)   Needle orientation diagram for the scene.

curved surfaces.   Sethi [20] has used characteristic contours, which correspond to equi–magnitude contours, to classify curved surfaces.   He used Hough transform to detect equi–magnitude contours.   In an earlier contribution, we proposed a surface classification scheme based on equi–magnitude and equi–orientation contours; these contours were constructed with the aid of needle code histograms [26] — the needle code histogram can be considered as a two–dimensional representation of the EGI (Extended Gaussian Image) [8,9].

9.3.4   SYNTHETIC CONTRAST GENERATION FROM NEEDLE ORIENTATION AND MAGNITUDE DATA

When a viewer perceives a surface, its apparent brightness depends upon a number of factors, amongst them being the light source direction, the viewer direction, the surface reflectance property, the surface orientation, etc. If all other factors are held constant, the perceived brightness of an object surface depends primarily on its orientation.

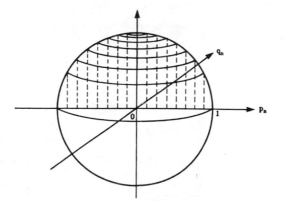

Figure 24(a) Depicted here are circular bands on a Gaussian sphere that are spanned by different bins whose boundaries are given by uniform quantization of the magnitudes in the (p_n, q_n) space.

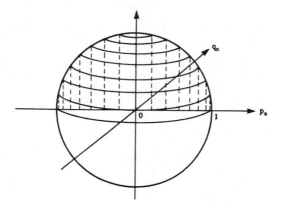

Figure 24(b) With sinusoidal quantization of magnitudes in the (p_n, q_n) space, projected regions on the Gaussian sphere are more nearly equal.

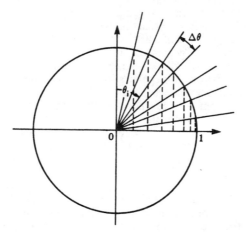

Figure 24(c) Sinusoidal quantization of needle magnitudes is depicted here by viewing the Gaussian sphere along the q_n axis.

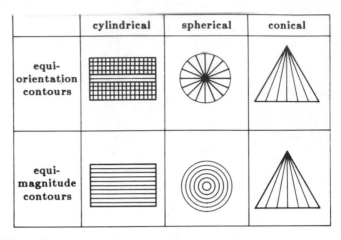

	cylindrical	spherical	conical
equi-orientation contours			
equi-magnitude contours			

Figure 25   Equi–orientation and equi–magnitude contours for three different curved surfaces.

Since a needle orientation diagram is a viewer centered representation of local surface orientations, it is possible to generate a shaded 2–D image of a 3–D object with the gray level for each surface element depending locally on the needle orientation at that site.   Synthetic contrast generated in this manner can then be used for the detection of object edges, and this can be accomplished by using the usual image processing edge detection operators — in much of our own work we have used the Sobel operator.   Purely for numerical convenience, we make the gray level value assigned to a surface element equal to the quantized value of the needle orientation there; this is done after the quantized values are expanded from the range 0–31 to 0–255.

It is easily demonstrated by counter–example that the above strategy for the detecting of object edges must be supplemented by needle magnitude information.   Surface orientation in 3–D space has two degrees of freedom, as, for example, represented by p_n and q_n, and, therefore, cannot be represented by a needle orientation alone.   What is being said here is that the surfaces of many orientations in 3–D will have the same value of $\tan^{-1}(q_n/p_n)$, and therefore synthetic contrast generated by needle orientations alone does not suffice for the detection of object edges.   Similarly, if we only used the needle magnitude information for generating a synthetic contrast image, it also would not bring out all the object edges. In Fig. 26a we have shown a sketch of a pyramid viewed from above; the needle magnitude codes for each of the triangular surfaces will be identical, causing all four edges shown in the figure to be invisible in a synthetic contrast image generated from the needle magnitude data.   In Fig. 26b is shown a sketch of an octagonal prism viewed from above; its edges marked 1 and 3 will not be discerned in a synthetic contrast image generated from just the orientation codes.

Fortunately, if we process separately the synthetic contrast images corresponding to needle orientations and magnitudes, and then if we combine the edges obtained from each, we get all the object edges.   When we say all the object edges, we mean all the edges in the part of the object that is visible from where the range map was constructed.

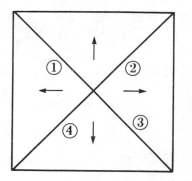

Figure 26(a)   For a pyramid viewed from the top, the needle magnitude codes for the four triangular surfaces are identical.

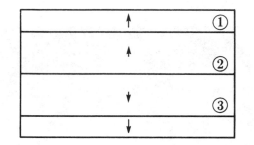

Figure 26(b)   For an octagonal prism viewed from the side, the needle orientation codes of the surfaces that join either at edge 1 or at edge 3 are identical.

Fig. 27a illustrates a synthetic gray level image generated from Fig. 22b, and Fig. 27b from Fig. 23b.

In a later section, we will show object edges extracted by this procedure.   As the reader will perhaps appreciate, this approach to the extraction of object edges is far superior to the best that can be done by applying edge detection algorithms to photometric information.

9.3.5   DATA BANDWIDTHS AND COMPUTATIONAL COMPLEXITY

Let's assume a structured light system capable of collecting 256×256 range maps in one second.   Note that range mapping consists of only recording the locations of pixel offsets in each scan line of a camera.   We record the offsets with 8 bit precision; this leads to a data bandwidth of $256\times256\times8 = 0.5$ Megabaud.

If we convert a 256×256 range map into a 100×100 needle diagram, the total number of arithmetic operations would be approximately 4×10^5, since it takes about 40 real arithmetic operations to fit a planar patch to an 8×8 set of pixels.   An arithmetic operation here can either be an addition or a multiplication.   Therefore, on a fast chip like TMS320 with a multiply time of 200 ns, it would take less than a second to compute the needle diagram from

Figure 27(a)   Synthetic gray level image constructed from the needle diagram in Fig. 22b.

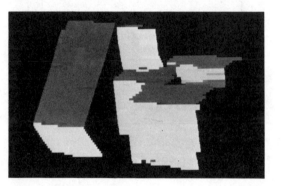

Figure 27(b)   Synthetic gray level image for the needle diagram in Fig. 23b.

the range map.   Also, the computation of the needle diagram can be conducted almost simultaneously with the recording of the range map.

For the spatial resolution used in the results shown in this paper, the computations take even less time.   Our range maps are 480×100 (for 100 light stripes across the field of view) and we convert these into 120×50 surface patches.

9.4   EDGE EXTRACTION AND LABELLING BASED ON SYNTHETIC GRAY LEVEL IMAGE AND OFFSET DATA

9.4.1   EDGE CATEGORIES

We categorize edges into one of: convex, concave, and obscuring.   A convex edge is where two surfaces of an object recede away with increasing distance from the observer (Fig. 28).   At a concave edge, as is evident from the same figure, the two adjoining surfaces face each other.   Obscuring edge is where range data has a jump discontinuity.

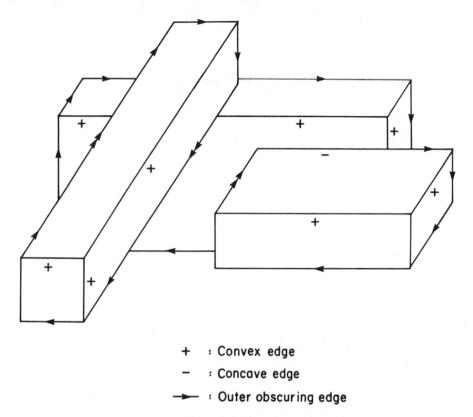

+   : Convex edge

−   : Concave edge

→► : Outer obscuring edge

→►► : Interior obscuring edge

Figure 28   Convex, concave and obscuring edges.

There are two types of obscuring edges: edges where range discontinuities are from object points on one side to background points on the other; and edges where the range discontinuities are between different objects in the scene or between different parts of the same object. The latter type of obscuring edges are called *interior obscuring edges*; these will be shown to play a very important role in scene segmentation.

Although we do not use them here, it is also possible to use other edge labels for scene analysis. For example, Waltz [24,25] has used crack and shadow edges. As shown in Fig. 29a, a crack edge occurs where one object abuts another, without a gap between the two. Shadow edges, as the name implies, are those edges that delineate the boundary of a shadow in a scene. Clearly, shadow edges are not pertinent when processing range maps. Sugihara has defined obscured edges (Fig. 29b) that denote boundaries between the visible and the obscured parts of an object surface [22]. Bolles [3] categorized obscuring edges into tangential, jump and sheet edges [3]. As depicted in Fig. 29c, a tangential edge is formed by a surface curving away from the viewer. The differences between a jump edge and a sheet edge is

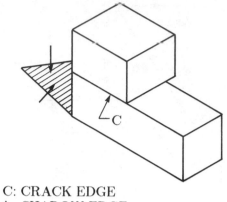

C: CRACK EDGE
↑ : SHADOW EDGE

Figure 29(a)   Crack and shadow edges defined by Waltz.

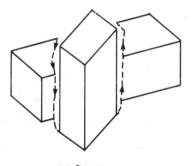

OBSCURED EDGE

Figure 29(b)   Obscured edges as defined by Sugihara.

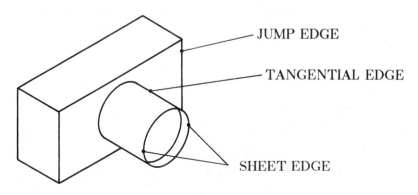

JUMP EDGE

TANGENTIAL EDGE

SHEET EDGE

Figure 29(c)   Jump, tangential and sheet edges as defined by Bolles.

determined by how far the observed surface would have to be folded so that it cannot be seen. The purpose for defining a sheet edge is to try to distinguish thick volumes from thin ones.

Our edge extraction and labelling scheme was developed for segmenting scenes consisting of piles of polyhedral objects using edge–vertex line drawings obtained from range map [4]. For this type of processing, described in detail in [4], it is sufficient to label edges as convex, concave, outer obscuring and interior obscuring.

9.4.2   CONSTRUCTION OF THINNED, 8–CONNECTED AND LABELLED EDGE MAPS

A schematic diagram depicting the procedures for our edge extraction and labelling scheme is shown in Fig. 30. In the diagram, the input is raw needle orientation and magnitude data stored in the light source plane and the output is a thinned, 8–connected, and labelled edge map. The output is also referred to as a *Line Drawing*, since any linearly connected set of edge labels will be a line representation of an edge in the 3–D scene. Following labels are used in the output:

    0:   Background Region
    1:   Object Region
    2:   Convex Edge
    3:   Concave Edge
    4:   Outer Obscuring Edge
    5:   Interior Obscuring Edge

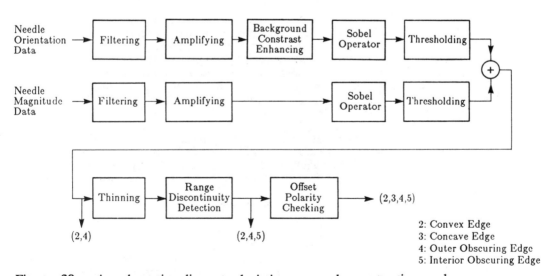

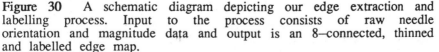

Figure 30   A schematic diagram depicting our edge extraction and labelling process. Input to the process consists of raw needle orientation and magnitude data and output is an 8–connected, thinned and labelled edge map.

For most experimental situations, the raw needle magnitude and orientation data needs to be "cleaned up" before it is converted into synthetic gray level images. The following factors are responsible for any noise that may be present in needle magnitude and orientation values:

(i)   Thresholding and thinning errors in stripe detection,

(ii)   8×8 patches located on surfaces highly oblique with respect to projected beams,

(iii)   8×8 patches straddling two adjacent surfaces of different orientations on the same object,

(iv)   8×8 patches straddling two different surfaces at different depths, meaning that there is a range discontinuity within the patch area.

Thresholding and thinning errors in stripe detection are caused by uneven reflectance of object surfaces and by uneven focusing of the stripe projector. Errors generated by patches located on surfaces highly oblique with respect to the projected beams are caused by two different factors: First, as a surface becomes increasingly oblique with respect to a projected light stripe, it causes the stripe to become smeared and distorted — smeared and distorted stripes exacerbate problems with stripe detection and thinning. (Note that errors in stripe detection and thinning translate into errors in the computation of the coordinates of illuminated points). Second, the physical surface area covered by an 8×8 patch increases as the obliquity angle of a surface with respect to projected light beams gets larger; it is possible that for larger areas, a planar fit may be a poor approximation to the local surface shape.

The situation when a patch straddles two adjoining surfaces of different orientations can be detected, when the orientations are considerably different, by thresholding the mean square error between the fitted planar patch and the actual locations of the illuminated points in the patch. However, the selection of a threshold for this purpose depends upon the accuracies in stripe detection, thinning, and a host of other factors; and, no matter where the threshold is placed, it would allow those patches to go through that straddle adjoining surfaces with nearly similar orientations.

The raw needle magnitude and orientation data is mostly free from the errors caused by patches referred to in item (iv) above. As mentioned in Section 9.3.2, when we calculate local surface normals by the least squares method, we use only those points whose pixel offsets are within a certain threshold from that of the center point of the patch. This precludes acceptance of patches over range discontinuities.

To rid the needle magnitude and orientation data of some of these errors, we first apply a median filter to the data; this process consists of replacing the magnitude and orientation values for a patch by medians computed for a 3×3 matrix of patches containing the patch in question at its center. In some cases, not all nine patches in a 3×3 set might be acceptable for the computation of the medians; some may be rejected if their pixel offsets are too far (meaning that they are probably on a different surface) from that of the patch at the center.

After the raw needle orientation and magnitude data is cleaned up by median filtering, we amplify the magnitude and orientation codes to cover a gray level range of 0–255. It is then possible to directly assign 8–bit gray level values to surface elements in order to construct synthetic gray level images corresponding to needle magnitudes and orientations. At this point, we also assign a large value, 1000, to the background for the purpose of enhancing contrast between the object region and background in the synthetic images.

To the synthetic images thus produced, any typical gray level edge detector can be applied for the determination of object edges. We have used Sobel operator for this purpose. Since the output of the Sobel operator at the outer obscuring edges is quite distinguishable from the output at the interior edge points, it is a rather simple matter to select a threshold for locating outer obscuring edges. We use the label '4' for labelling such edges; all the other edges detected by the Sobel operator are at this time labelled as '2'. We label outer obscuring edges as '4' and temporarily label the rest of the edges as '2'; eventually, the label '2' is only used for convex edges.

The symbolic processing programs that we use for scene segmentation require that the object edges extracted from our range maps be thin (one element wide, if possible) and 8–connected. For this purpose we use the thinning procedure described in [19], which consists of stripping away points that are called 8–simple. To ensure that a thinned edge lies as close as possible to the middle of the edge that existed before thinning, stripping is done alternately from the four cardinal directions.

The processing described so far generates an edge map of the scene with outer obscuring edges clearly identified. The rest of the edges, which at this time all have the label '2' must now be categorized as interior obscuring, convex or concave.

Interior obscuring edges are located by detecting range discontinuities inside the object region (i.e., those points that are not labeled as background points) and for this purpose the offset data can be used directly, since all the range discontinuities are reflected in this data. We label all such edge points with '5'. Note that a range (or offset) discontinuity inside the object region may manifest itself in two different ways: A discontinuity may occur between two adjacent light stripes; or it may occur between two rows in the light source plane. As shown in Fig. 31, the former case is caused by an object surface that may be visible from the camera viewpoint, but is hidden from the light source viewpoint. As is clear from the figure, such an edge could be declared as convex, since that is its nature from the camera viewpoint, but we do not do so. When a range discontinuity is of the latter type, it can only be called interior obscuring.

Finally, we must discriminate between convex and concave edges; this we do by using an operator first suggested by Sugihara [22]. As shown in Fig. 32, suppose we have a concave edge that is nearly horizontal, the following second difference operator applied to the offset data along a stripe would always be negative:

$$P_h(i,j) = [d(i-k,j) + d(i+k,j) - 2d(i,j)]/2k \quad .$$

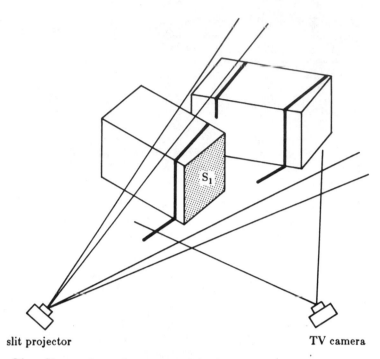

slit projector   TV camera

Figure 31   Shown here is an example in which a range discontinuity between two adjacent light stripes is caused by an object surface that is visible from the camera viewpoint, but is hidden from the light source viewpoint.

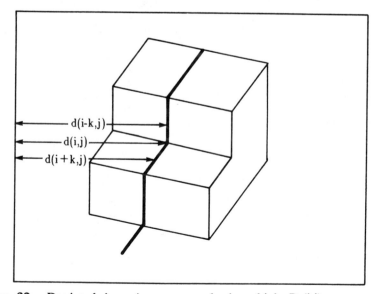

Figure 32   Depicted here is an example in which $P_h(i,j)$ at a concave edge, which is nearly horizontal, has a negative value.

For a convex edge that is nearly horizontal, P_h would be positive. Since this second order difference is obviously sensitive to the orientation of a convex or concave edge, we also evaluate the following expression at each edge point; this expression has sensitivities orthogonal to the one for P_h:

$$P_v(i,j) = [d(i,j-k) + d(i,j+k) - 2d(i,j)]/2k \quad .$$

For a convex edge that is nearly vertical, P_v will always be positive, and negative if the edge is concave. The value to be used for k depends upon the resolution of the light source image along and perpendicular to the light stripes. We have used k=5 for P_h and k=2 for P_v. Since P_h and P_v have orthogonal sensitivities, we declare an edge point to be either convex or concave depending upon the sign of the one that has a larger magnitude.

In Figs. 33a and b are shown, respectively, 8–connected and thinned edges of the 2–box scene of Fig. 9a and the 5–box scene of Fig. 23a. An 8–connected edge in the light source may lose its continuous appearance in the camera image plane, especially if, for a surface of high obliquity, the stripes are far separated. This is the reason why the edges look rather jagged in the figures. By increasing the number of stripes, the jagged appearance can be eliminated. Although, it is not possible to show them, all the edges in Fig. 33 have labels that were computed by the method described here.

Edge maps, such as those shown in Fig. 33, are used by our pile analysis procedure discussed in [4] to segment out objects from a scene. Fig. 34 shows an example of such segmentation.

9.5 CONCLUDING REMARKS

We have found 3–D vision via structured light to be a rugged sensing modality for integration with manipulation. It is not that difficult to design light–weight structured light sensors that can either be mounted on a robot

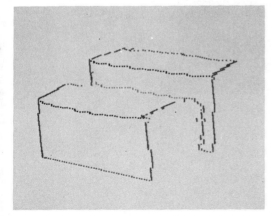

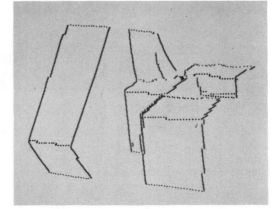

Figure 33(a) Detected edges in the image of Fig. 27a.

Figure 33(b) Detected edges in the image of Fig. 27b.

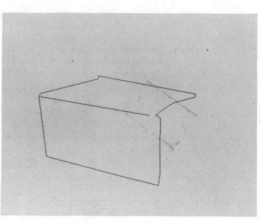

Figure 34(a)   Edges and vertices corresponding to the front box have been segmented from the edge map shown in Fig. 31a.

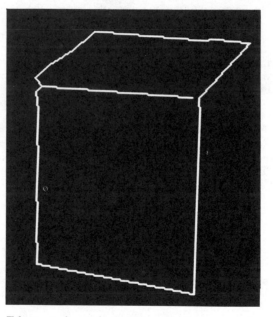

Figure 34(b)   Edges and vertices corresponding to a box in Fig. 31b have been segmented out

arm, or can be housed in a cradle, to be picked up by the robot when it needs 3–D vision information for interacting with a scene.   As we have shown here and, also, as demonstrated earlier by Sugihara [22], a good deal of scene interpretation can be carried out directly by processing the offset information without converting it to range maps, which leads to great savings in computations.

Our emphasis in this paper has been on different structured light methods for generating range maps of scenes, and on how to carry out scene analysis by extracting and labelling edges and then how to label them. In many cases, superior scene analysis strategies can be devised by basing them on region–growing concepts; the reader is referred to [17,27,28] for further discussion on such methods.

9.6   ACKNOWLEDGEMENT

This work was supported by the Natural Science Foundation and by the Purdue Engineering Research Center for Intelligent Manufacturing.

9.7   REFERENCES

1.   G. Agin and T. Binford, "Computer Description of Curved Objects," *3rd IJCAI*, 1973, pp. 629–640.

2.   D.H. Ballard and C.M. Brown, Computer Vision, Prentice–Hall, Inc. N.J., 1982.

3.   R.Bolles, "Three–Dimensional Locating of Industrial Parts,"

4.   K.L. Boyer, H.S. Yang and A.C. Kak, "3–D Vision for Pile Analysis," Purdue University Technical Report, TR–EE–84–17, 1984.

5.   K.L. Boyer and A.C. Kak, "Color–Encoded Structured Light for Rapid Range Map Computation," To appear in *IEEE Trans. PAMI*, Sept. 86.

6.   R.O. Duda, D. Nitzan and P. Barret, "Use of Range and Reflectance Data to Find Planar Surface Region," *IEEE Tran. PAMI*, 1979, pp. 259–271.

7.   B. Gil, A. Mitiche and J.K. Aggarwal, "Experiments in Combining Intensity and Range Edge Maps," *CVGIP* 21, 1983, pp. 395–411.

8.   B.K.P. Horn and K. Ikeuchi, "The Mechanical Manipulation of Randomly Oriented Parts," *Scientific American*, 1984, pp. 100–111.

9.   B.K.P. Horn, "Extended Gaussian Images," *Proceedings IEEE*, Dec. 1984, pp. 1671–1686.

10.   S. Inokuchi and R. Nevatia, "Boundary Detection in Range Pictures," *ICPR*, 1980, pp. 1301–1303.

11.   S. Inokuchi, T. Nita, F. Matsuda and Y. Sakurai, "A Three Dimensional Edge Region Operator for Range Pictures," *ICPR*, 1982, pp. 918–920.

12.   S. Inokuchi, K. Sato, and F. Matsuda, "Range–Imaging System for 3–D Object Recognition," *Proc. 7th ICPR*, 1984, Vol. 2, pp. 806–808.

13.   A.C. Kak, "Depth Perception for Robots" in *Handbook of Industrial Robotics*, S. Nof (ed.), John Wiley, New York, 1985, pp. 272–319.

14.   A.C. Kak and J.S. Albus, "Sensors for Intelligent Robots," in *Handbook of Industrial Robotics*, S. Nof (ed.), John Wiley, New York, 1985, pp. 214–230.

15.   M. Minou, "Theoretical and Experimental Studies on Basic Relations between Real World Pictorial Patterns and their Generating Constraints," Ph.D. Dissertation, Nov. 1982, Dept. of Information Science, Kyoto University.

16.   D. Nitzan, A.E. Brain and R.O.Duda, "The Measurement and Use of Registered Reflectance and Range Data in Scene Analysis," *Proc. IEEE*, 1977, Vol. 65, pp. 206–220.

17.   M. Oshima and Y. Shirai, "Object Recognition Using Three Dimensional Information," *IEEE PAMI*, Vol. PAMI–5, July 1983, pp. 353–361.

18.   F. Rocker and A. Kiessling, "Methods for Analyzing Three Dimensional Scenes," *Proc., 4th IJCAI*, 1975, pp. 669–673.

19.   A. Rosenfeld and A.C. Kak, Digital Picture Processing, Vol. 2, 2nd ed., Academic Press, pp. 232–240.

20.   I.K. Sethi and S.N. Jayaramamurthy, "Surface Classification using Characteristic Contours," *7th ICPR*, 1984, pp. 438–440.

21.   Y. Shirai, "Recognition of Polyhedrons with a Range Finder," *Pattern Recognition*, Vol. 4, 1972, pp. 243–250.

22.   K. Sugihara, "Range Data Analysis Guided by a Junction Dictionary," *Artificial Intelligence*, 1979, pp. 41–69.

23.   F. Tomita and T. Kanade, "A 3–D Vision System: Generating and Matching Shape Description in Range Images," *1st Conference on Artificial Intelligence Applications*, Dec. 1984, pp.186–191.

24.   D.L. Waltz, "Generating Semantic Descriptions from Drawings of Scenes with Shadows," MIT Report AI–TR–271, Nov. 1972.

25.   D.L. Waltz, "Understanding Line Drawings of Scenes with Shadows." in *The Psychology of Computer Vision*, P.H. Winston (ed.), McGraw–Hill, New York, 1975, pp. 19–91.

26.   H.S. Yang, K.L. Boyer and A.C. Kak, "Range Data Extraction and Interpretation by Structured light," *1st Conference on Artificial Intelligence Applications*, Dec. 1984, pp. 199–205.

27.   H.S. Yang and A.C. Kak, "Determination of the Identity, Position and Orientation of the Topmost Object in a Pile: Some Further Experiments" *Proc. IEEE International Conference on Robotics and Automation*, 1986, pp. 292–298.

28.   H.S. Yang and A.C. Kak, "Determination of the Identity, Position and Orientation of the Topmost Object in a Pile", *Computer Vision, Graphics and Image Processing*, Vol. 36, Nov. 1986, pp. 229–255.

10
BIOMEDICAL SIGNAL PROCESSING

D.G. CHILDERS
Department of Electrical Engineering
University of Florida
Gainesville, FLA

10.1 INTRODUCTION

This paper opens with a survey of the origins, costs, implications, and directions of biomedical signal processing.   While the primary theme is the processing of signals from the brain, such as electroencephalography (EEG) and event–related potentials (ERPs), the techniques are applicable to other signals as well.   The procedures discussed cover simple signal averaging, signal conditioning, filtering, spectral analysis, feature extraction , pattern recognition, adaptive algorithms, and the application of artificial intelligence techniques. Analytic detail is provided along with discussion of applications.   While a major aspect of biomedical signal processing is the monitoring, analysis, interpretation, and validation of signals, a new emerging area is the development of neural control devices including human–computer interfaces that make use of body signals, speech, brain waves, pointing, touching, and eye movements to facilitate human–computer interaction for the control of machines and devices.   We conclude with an overview of two of the newest research areas, molecular computing and neural networking.   Molecular computing will use molecules to sense, transform, compute and provide output signals.   Neural networks model brain functioning and will influence new computer architectures. The chapter interprets biomedical signal processing as a broad, exciting field for those who enjoy applying technical skills to solve detective problems in the life sciences and assist the clinical diagnosis of diseases.

10.2 TECHNOLOGY AND BIOMEDICAL SYSTEMS

10.2.1 SCOPE

Biomedical signal processing has two major application areas.   One is to aid the scientific investigation of the functioning of biological systems.   The other is to assist the medical clinician, perhaps providing material useful to the diagnoses of a patient's medical disorder.   At the present time, other tasks are of lesser importance, such as the coding, storage, and communication of data. We look upon biomedical signal processing as an endeavor requiring exceptional breadth, demanding communicative, analytical, and experimental skills.   The field uses tools from communications, psychology, physiology, biology, pattern recognition, artificial intelligence, and electrical engineering. The problems to be solved require the use of computation, the development of algorithms, and the invention of new experimental designs.   The collection and organization of data is the heart of this activity.   The data are examined for clues to new or the validation of old hypotheses.   Data and hypotheses are the basis of new models to explain the functioning of biomedical systems.

But the field must also validate these models using new experiments and statistics, a task more important for this professional pursuit than in many other areas. In recent years the symbiosis of the human and the computer has stimulated new research for the design of human–machine systems. And molecular computing and neural networks are presenting exciting challenges for the biomedical engineer. Because biomedical signal processing is a servant of the scientist and the physician, there are issues other than just the technical aspects that must be considered. The reader might find these interesting.

10.2.2 ORIGINS

Prior to the beginning of the twentieth century the medical resources of this country were primarily the physician and the contents of his black bag. Physicians were in short supply but the demand for their services was also small. Friends, relatives, and neighbors helped nurse the ill. If a home remedy did not work, then the patient often died. Since 1900, the hospital has changed medicine. These complexes are technically sophisticated edifices with a highly trained staff. With the advent of the hospital, the physician's practice changed from treating infections to performing major surgery, including open heart surgery and organ transplants. These changes were driven by rapid advances at the turn of the century in chemistry, physics, engineering, microbiology, physiology, pharmacology, and others. William Enthoven invented the string galvanometer and used it to record the first electrocardiogram (ECG or EKG) in 1903, thus monitoring remotely the electrical activity of the beating heart. The electrocardiograph, adopted by some hospitals by 1910, lead the way for cardiovascular medicine and other biomedical electrical measurement methods [1].

Perhaps the most influential invention for clinical medicine was the develpment of x–rays. W.K. Roentgen said these new rays opened the inner man to medical inspection. By the 1930's the physician could see practically any organ of the body with the aid of barium salts and other substances. But x–ray machines were expensive and the hospital became the central place where such devices could be used by many physicians to assist in the diagnosis and treatment of numerous patients.

About the same time in 1929, Hans Berger first reported the human electroencephalogram, the recording of the brain's electrical activity. This procedure for diagnosing cerebral diseases also quickly became located in hospitals because of the cost involved.

We all think of spectrum analysis as a new tool, but 50 years ago, perhaps the first attempt to automate the analysis of recorded EEG data was that of Grass and Gibbs in 1938 [2]. They recorded the EEG on film and then repetitively played it back through an electronic filter (as one might play back the sound track of a movie). This gave them a frequency spectrum of the recorded data. Of course, today we use more modern techniques which we discuss later.

Early electronics contributed to the measurement of electrical behavior of the cells of the nervous system, the neurons. Nuclear medicine mushroomed with the atomic age. Diagnostic ultra sound techniques benefited from sonar developments. Intensive care units came into being. And more recently, we have seen the development of modern medical imaging devices such as the computerized axial tomography (CAT) scanner, nuclear magnetic resonance, digital radiography, and others.

The benefits of these advances include an increase in life expectancy from 48.2 years to 66.7 years for men between 1900 and 1950 and 51.1 years to 72.1 years for men and 78.3 for women. These increases have been due to better health care, nutrition and hygiene [1].

10.2.3 COSTS

But we have had to pay for these advances (See Table I). These increased costs are not due solely to expansions in technology since more people received basic health care when they were ill. But it is estimated that 75% of these increased costs are due to increased sophistication in medical procedures [1]. The other 25% is mainly attributed to salary differentials between health care prfessionals and other workers. Costs are perhaps more significant for biomedical signal processing than other technology areas because the public can more easily relate to health care and its sometimes life and death situations, and because health care costs have been debated extensively in the media in recent years.

TABLE I

HEALTH CARE COSTS [1]

YEAR	TOTAL EXPENDITURES BILLIONS OF DOLLARS	PER CAPITA EXPENDITURES DOLLARS	EXPENDITURES AS A PERCENT OF GROSS NATIONAL PRODUCT
1950	$ 52.4	$ 340	4.4%
1960	$ 90.5	$ 491	15.3%
1970	$183.6	$ 881	17.2%
1983	$355.4	$1499	10.8%

Technology is often accused of being a major contributor to the increase in the cost of medical care as well as to a decrease in the quality of patient life and a decrease in the humane aspects of medicine. Several studies have implied that 2 percent of the population use about 25 percent of all health care resources. Several high–cost technologies contribute heavily to cost increases despite their marginal benefit to many patients. These are coronary bypass procedures (e.g.: in 1982, about 125,000 were performed, costing 2 billion dollars), neonatal intensive units (e.g.: 6 percent of all live births, 200,000 per year, costing 1.5 billion dollars), and maintaining one brain damaged patient may cost $100,000 per year [3].

Clearly, only a portion of the previously mentioned technological advances and associated costs are directly related to biomedical signal processing. But even ten percent of the costs would be a large expense. And surely this percentage is too small.

10.3   BIOMEDICAL SIGNAL PROCESSING: EXAMPLES AND APPLICATIONS

The technological areas of biomedical signal processing are varied.   Let us look at a few.   Advances in biomedical signal processing and related areas in the 1950s saw the development of cardiac monitoring, catheterization, modern anesthesia, and the first computer for biomedical research [3].   The LINC computer, developed by W. Clark of M.I.T.'s Lincoln Laboratories in 1961 for biomedical research, brought about realtime, on–line experiments.   The computer was easy to program, easy to use, easy to maintain and could digitize and process biomedical signals directly.   This computer and those that followed revolutionized biomedical signal processing.

As we moved into the 1960s, we witnessed the development of intensive care units, scintillation cameras, implanted pacemakers, defibrillators, kidney dialysis, and ultra sound imaging.   Since computers were now available, the development of numerous data processing algorithms came about [4]. These first programs did not interpret computational results, but merely produced data from a formula such as calculating force given mass and acceleration.   By the 1970s we were seeing software systems for information processing and computer aided decision making, i.e., the organization of data into a meaningful format as in pattern recognition applications or a patient's medical record and history.   Knowledge engineering has transpired from the mid 70s into the 1980s.   Expert systems now attempt to formalize the rules or relations by which information is obtained from data.

The medical electronic market came into being.   This market may be divided into five major areas: diagnostic imaging, patient monitoring, therapeutic devices, medical laboratory instruments, and medical information systems. Revenues for these areas are projected to be 7.26 billion dollars in 1987 plus 3.97 billion for related products [3].

Diagnostic imaging has six areas: conventional x–rays, computed tomography, ultrasound, nuclear magnetic resonance, digital radiography, and nuclear medicine.   Research is needed for better qualification techniques, measurements of dynamic events, e.g., blood flow, and the assessment of long–term side effects.   We exclude this area from our survey of biomedical signal processing.

Patient monitoring has benefited from microprocessors and the decline in costs of computer technology.   Intensive care monitoring is still invasive, using 20 year old sensors, with no new noninvasive measurement methods on the horizon.

Therapeutic devices are, however, making advances toward techniques that are not unduly invasive.   These include biofeedback, pain control and bone tissue stimulation methods to promote bone growth as well as cardiac pacemakers and defibrillators.

Medical laboratory instrumentation is quite common and is perhaps most mature in automated chemistry analyzers.

Medical information and order communication systems include business machines and computer systems that cover the administrative and financial aspects of health care.

Another emerging area is the human–machine connection, which has the potential for the neural control of devices, including a computer, for various uses such as prosthetics and sensory aids and assists for other bodily functions. We might also have a simple "cortical switch" which is turned on or off by simply "thinking" a command. To achieve such goals, we need to monitor appropriate electromyograms (EMGs) (muscle potentials), electroencephalograms (EEGs), and micropotentials of individual nerve cells. These signals need to be studied for clues that correlate with thinking, memory, or the brain's initiation of a command to control the action of a muscle. Such knowledge is leading and will continue to lead to the development of methods to stimulate nerves and muscles.

Examples of existing devices or those undergoing research and development include the following. Pacemakers are used to assist or replace the heart's natural pacemaker cells. Commercial devices provide a signal that helps the heart initiate and coordinate each new heart contraction. There are now pacemakers to assist the activation of the diaphragm to initiate regular breathing in patients who have lost this ability. These devices function continuously, requiring no intervention from the patient and without the patient's awareness [5].

Growth devices involve the electrical stimulation of tissue or bone. Neuromuscular devices are being designed to strengthen weak muscles as in scoliosis (bent back) and strobismus (crossed eyes). Surgery and/or braces used to be the treatment. But this is changing as researchers are experimenting with chronic electrical stimulation often during sleep [5].

Central nervous system stimulation is being and has been experimented with to relieve pain and assist patients suffering from motor disorders, e.g., epilepsy, and spasticity [5].

Neural prosthetic devices interact with patients, often being under their control. Applications, besides the control of pain, include auditory prostheses, and control of the bladder. The electrodes used depend on the particular case but some have been implanted in the brain, spinal cord, peripheral nerves, and on the skin surface. Multiple implanted electrode arrays have been explored as a possible interface between a video camera and the brain for a simple visual prosthesis for the blind. Other devices have been marketed for the blind including walking and reading aids. The latter usually interface with the tactile system such as the fingers or even the skin of the stomach.

Neuromuscular stimulation is being consdered as an aid for the restoration of muscle control in paralyzed limbs as might occur in spinal cord injury or stroke. The evolution of these devices will require sophisticated multichannel stimulation under the regulated control of a computer with considerable computational power [5].

To achieve these advances, the researcher must reduce scientific data into meaningful indicators, probably the most important problem in biomedical signal procesing.   The competent physician is able to encompass only a small fraction of the new knowledge now available in medicine.   The brain seems able to manipulate less than ten diagnostic or therapeutic hypotheses at one time.   Consequently, unless new assistance is provided to help the clinician, he may not be able to reason properly about the appropriate information for the clinical problems at hand.   Knowledge–based expert systems may provide some help.   Perhaps MYCIN, the Stanford University system to help physicians decide the appropriate antibiotics to fight a patient's bacterial infection of the bloodstream, is the most renowned and also over a decade old.   An example of this system is [4]:

IF: 1) The site of the culture is blood, and
          2) There is significant disease associated with this occurrence
             of the organism, and
          3) The portal of entry of the organism is GI, and
          4) The patient is a compromised host

THEN: It is definite (1.0) that bacteroides is an organism which
        therapy should cover.

These systems apply hundreds of rules to the data supplied and make inferences and arrive at one or more conclusions.

A clinical decision making program using a frame–based, goal–directed approach was published in 1976.   This present illness program (PIP) used pattern matching and a large associative memory to diagnose edematous disorders, i.e., diseases of excessive fluid accumulation.   The program evolution modeled the physician's diagnostic behavior as well as the pathophysiologic knowledge of the disease.   The object was to assemble many small problem–solving strategies.   The successful solution of these many sub–problems yielded the correct diagnosis.   Diagnostic hypotheses were generated to assist the information–gathering process.   As new information was gained the hypotheses were scored to determine if they should be rejected, accepted or considered further [4].

Having covered the scope, history, costs, and a few of the areas of biomedical signal processing, let us now look more closely at some typical signals and their characteristics and the sensors that are used to monitor these signals.

10.4   SIGNALS AND SENSORS

Biomedical signals are quite varied in their bandwidth and duration.   The signals monitored and the features selected are frequently determined by the application.   A partial listing of signals appear in Table II.   The bandwidths vary from nearly DC for temperature to 50 Hz for ECG to 3000 Hz for EMG to 5000 Hz for speech.   The imaging signal bandwidths are larger yet.

Following World War I, the triode amplifier became available. Scientists could now record biopotentials and display them on the cathode ray oscillograph (Braun tube) which had been invented in 1897.   With the aid of Western Electric Engineers, Erlanger and Gasser [6] were able for the first

TABLE II

BIOMEDICAL SIGNALS

BIOMAGNETISM
        MAGNETOENCEPHALOGRAM (MEG)

CARDIAC
        ELECTROCARDIOGRAM (ECG OR EKG)

CHEMICAL
        PH
        URINE
        BLOOD OXYGEN
        ETC.

CORTICAL
        CORTEX
        SCALP
        ELECTROENCEPHALOGRAM (EEG)
        EVENT RELATED POTENTIAL (ERP)

DIAGNOSTIC IMAGING
        COMPUTERIZED AXIAL TOMOGRAPHY (CAT)
        NUCLEAR MAGNETIC RESONANCE (NMR)
        POSITRON EMISSION TOMOGRAPY (PET)
        ULTRASOUND
        X–RAYS
        DIGITAL RADIOGRAPHY
        NUCLEAR MEDICINE

LUNG SOUNDS

MUSCLE
        ELECTROMYOGRAM (EMG)

NEURAL
        ACTION POTENTIALS

SPEECH

TEMPERATURE
        THERMAL MAPPING OF THE BRAIN'S METABOLIC
        RATES

TISSUE IMPEDANCE

time to record the shape of the action potential. Their analysis showed that
this potential was the sum of the potentials from fibers of differing sizes and
types with different conduction velocities. They were awarded the Nobel Prize
in 1944 for this work.

Electrical potentials originate in biological systems (amoeba, plant or human) as a result of concentration gradients of ions across the membranes that surround the cells of which the organism is composed. In general, these membranes tend to let some species of ions pass readily, but offer resistance to others [6]. One tenth of a volt is about the highest potential across a single membrane and is the driving mechanism for most biological signals. The transmission of these signals has been studied since Galvani in 1760. Hodgkin and Huxley worked on this problem starting in 1952. They too were later awarded the Nobel Prize in 1963, which they shared with Eccles.

The sensors used to monitor biopotentials vary. But the most common sensor is the electrode, the materials of which also vary, including gold, silver, silver—silver—chloride and others. Some electrodes are invasive, piercing the cortex or a muscle. Others are applied to the skin, sometimes with the use of an electrode paste to facilitate the connection between the skin's surface and the electrode interface. Arrays of electrodes are frequently used to effect a two— or three—dimensional spatial sampling of the signal space.

A skin or scalp electrode averages the signal it receives over its surface area. The signal received is dependent on the electrode size and is a composite of potentials close to and far from the recording site. There are multiple pathways by which the signal may reach the recording sites. The electrode montage and interconnections (monopolar or bipolar) also influence our interpretation.

One example of a surface potential is the electrocardiogram (ECG), which records the cardiac activity through a series of electrodes placed on the body. Standards have been adopted so that physicians throughout the country can interpret a patient's EGC regardless of the manufacturer of the device, who recorded it or where it was recorded. Each segment of the ECG has physiological significance (see Figure 1). The P wave corresponds to the depolarization of the atria. Next, the QRS component comes about from the depolarization of the ventricles. Finally, the T wave segment represents the repolarization of the ventricles. The heart's electrical activity is controlled by

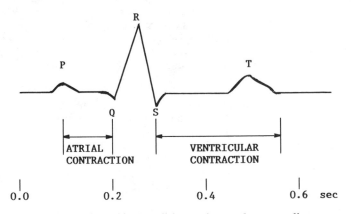

Figure 1 Schematic rendition of an electrocardiogram.

its own pacemaker cells known as the sinoatrial (SA) node. Physicians have learned to recognize the effects of various heart disorders on ECG segments. If the sensing electrodes are shifted in position, then different areas of the heart can be analyzed, even specific chambers within the heart. In this way, the ECG can be used to diagnose the death of a portion of the ventricle muscle from a heart attack [7].

In recent years, technology has provided the means to measure the magnetic fields associated with the function of biological systems [39–41]. The superconducting quantum interference device (SQUID) sensor is at the centre of this technology. Magnetic fields are not greatly distorted by the intervening biological tissue and the magnetic flux is measured perpendicular to the skin's surface. The SQUID's sensors do not contact the skin's surface but are placed in close proximity to the area to be measured. Research is underway examining the magnetic fields associated with the human brain, heart, lung, liver and other tissues. The technique is still experimental and the equipment is very costly, but it is possible to map magnetic field patterns directly onto photographs of the patient's head, for example. We shall describe the brain's magnetic fields in the next section.

10.5 THE BRAIN AND ITS POTENTIALS

The nervous system, with its sensors (temperature, tactile, visual, auditory, pain, and others) is our interface to the environment, providing information through our senses relevant to controlling our movements, our breathing, heart rate, and even our digestion. The brain controls body growth, development and reproduction. The nervous system with the exception of the most peripheral nerves does not grow, renew or repair itself as every other body tissue does. We can imagine the brain as a computer whose hardware is essentially fixed from birth but whose software is continually changing throughout life as we learn from new experiences. (Neural network modeling is changing this simplistic view).

The neurobiologist tries to understand the brain's "hardware" such as nerve cells, their interconnections, hierarchy, and processing. They hope to discover among other things, the physical nature of change in the brain during learning. The experimental psychologist is concerned less with how individual brain components work, but more with the "system" or "logic" of how the components are put together to produce the observed behavior of the whole organism.

The brain is self–maintaining, consuming 10 to 25 watts in a volume of 1500 cm^3. The rate of information transfer is slow, being 10 to 30 bits per second. However, the processing is both serial and parallel involving large groups of cells, and is much faster in some situations than any existing computer e.g., pattern recognition. There are essentially no errors in the brain's calculations, except in pathological cases. The reliability is high because of built–in redundancy in the brain.

It has been just over a century since Dr. Richard Caton, an English physician, reported in 1875 his observations concerning brain waves recorded from the exposed surface of the brain of rabbits and monkeys. He was the first to observe two forms of brain electrical activity: the first is now known as an evoked potential; and the second is the spontaneous, on–going electrical activity, which is now termed the EEG.   An evoked potential is the electrical activity at the brain's surface evoked by a sensory or event related stimulus and is usually measured over the sensory region of the brain corresponding to the sensory modality being stimulated e.g., the occipital region for a visual stimulus.   The event–related potential is known as an ERP and is measured from the scalp as is the EEG [8].

At the time of Caton's discovery it had been known for nearly one–half century that nerves conducted electrical pulses.   Another one–half century was to pass following Caton's investigations before brain waves were to be discovered in humans by the German psychiatrist, Hans Berger in 1929. By this time the technology had developed sufficiently beyond the crude galvanometer and optical amplification of Caton's day to the use of electronic amplifiers to enhance the sensitivity of the string galvanometer.   These developments allowed Berger to measure potential fluctuations directly from the scalp with two large pad electrodes soaked in saline which he placed on the forehead and at the back of the head over the occipital region.   As time passed, he confirmed his observations at the scalp by measuring directly from the surface of the brain, where he noted that this potential was larger in amplitude than the corresponding scalp potential.

What is the electrophysiological origin of these brain waves?   Our present understanding of nerve spike generation and propagation, synaptic potentials, and their relationship to the EEG, can be briefly described as follows.   Our brain is connected to the outside world through our sensory organs which act as transducers converting physical energy, such as auditory pressure waves or light waves, into a series of pulses which are conveyed along nerve fibers to the brain.   The intensity or amplitude of the sensory stimulus is encoded or represented as the pulse repetition frequency.   The brain not only responds to sensory stimuli, but also "thinks".   Both activities involve cognitive processing at the neuronal level; thus both are reflected in the EEG.

The nerve cell is a membrane that separates two media, the cytoplasm within the cell body and the extracellular fluid.   The ionic concentrations are considerably different between these two media.   The membrane is normally polarized at a potential difference of approximately 70 to 100 mV with the inside negative relative to the outside.

A nerve cell has a dendritic field (many small fibers) that feed graded slow potentials to the cell body.   These potentials are either depolarizing (exciting) or hyperpolarizing (inhibiting).   Their collective effect increases or decreases respectively the probability of the cell discharging (firing) a spike along its axon to the next cell.

The axon is the interconnecting cable between cells, but this cable is not passive, instead the pulse is regenerated along the axon analogous to repeaters in telephone lines.   Conduction continues along the axon to synaptic junctions at either other cell bodies or the dendrites of other neurons.   These synaptic junctions, which may be as numerous as 1000 per axon, conduct in only one direction.   When the nerve impulse arrives at one of these junctions, it triggers the release of a chemical transmitter that ferries the effect of the nerve pulse across the junction by chemicals. The chemicals in turn produce graded slow potentials called post synaptic potentials (PSPs), which may be as large as 30 mV, in the recipient cell and its dendrites. As noted above, the potentials may be either excitatory (EPSP) or inhibitory (IPSP). Thus, we may draw the analogy that the synapse is a digital–to–analog converter (D/A); the cell is an A/D, and the axon transmits spikes or digital signals with repeaters along the route.

The origin of the EEG has been explained in many ways. One of the first such attempts suggested that the brain wave was the envelope of activity resulting from the summation of action potentials or nerve spikes originating in cohesive neuronal ensembles. This hypothesis considered the extracellular space to serve not only as a conductive and dispersive medium, but also required some form of non–linear transformation such as a square–law device to rectify the negative components of the action potentials. Otherwise, cancellation and interference would occur between the positive and negative components of the nerve spikes. The medium then served as a low–pass filter, the output of which was the envelope of activity or brain wave. This conjecture soon gave way to the supposition that slow (brain) wave activity was the result of the summation of slower dendrite spikes and, further, it was also suggested that the spike after–potentials contributed significantly to the EEG. Then it was demonstrated that spikes could actually be abolished without affecting EEG. This led to the conjecture that the EEG and evoked potential must be produced by the post synaptic potentials since the fluctuations in PSP's correlate well with the cortical EEG. This latter hypothesis is presently in vogue with most researchers, although it too is undergoing modification, e.g., some feel there is evidence that the EEG is produced by a very large number of generators of cellular dimensions, which may be independent or even possibly nonlinearly related, and yet others feel that the EEG and ERP include contributions from both spike and PSP sources. This latter case has some possibility, even though nerve spike fields are smaller than PSP fields, i.e., PSP's can be seen at a greater distance; also multineuronal studies have shown that relationships between certain nerve spike patterns and gross (slow–wave) responses do exist. Despite the fact that the origin of brain waves is still enigmatic, it has been possible to develop important and useful correlations between the EEG and ERP and basic cerebral processes.

It is safe to say that the electrical activity monitored at one scalp location is a complex, intricate waveform whose components may reflect changes in the cerebral location of its source or sources and/or perhaps reflect changes in the temporal activity in the generative mechanism of a particular fixed source. Are these waves we monitor electrical signatures of neurochemical reactions? Are they indicators of information transfer between neuronal populations? Can they be used as reliable diagnostic aids? These, of course, are the "grand" questions we seek answers for, but for which, unfortunately, no "grand" answers presently exist.

Instead, investigators have been working on a more rudimentary level. Basically, two models have been advanced to explain cerebral functioning as seen through brain waves. One is the field theory approach and the other is the complex cortical connection theory. The former studies volume conduction while the latter considers potentials as they propagate along specific pathways. The electric field has been discounted by some as a means by which information is transferred in the nervous system but it is still used to model evoked response and EEG potentials. Here dipoles are typically assumed to be the sources of these potentials. But gross dipole models frequently lack credibility since it is doubtful that a large dipole (or a combination of large dipoles) covering an entire cortical region or some large portion of it could exist underneath the scalp; further, when some investigators compared their experimental data and mathematical results, they were led to conclude that the evoked response did not seem to be produced by a single dipole–like field. Others, however, have had good success along these lines.

Certain portions of the cortex do apparently display dipole–like properties. These dipoles are not of the gross variety that span an entire cortical area, but rather have a microscopic nature, although some feel the situation is considerably more complex than this. Since brain potentials are basically caused by ions moving across membranes, then dipole models are probably not correct or are at least an oversimplification. This is because these models are typically used to mathematically describe an isotropic medium, which brain tissue certainly is not.

Theories which embody both microdipoles and the concept of complex interconnections have been developed and show promise in modeling ERPs, but little or no effort has been expended to relate such models to complex grain structures; this is also true of the generator models. In the area of vision, some work has been done to devise a model to relate surface potentials to neuronal functioning.

In truth, the situation is generally such that no general model or theory exists to account for data beyond that of a particular investigator's experimental evidence. No correlations exist, sometimes strong and sometimes weak, between components of the EEG or ERP and sensory stimulus parameters. But a predictive model relating central nervous system physiology and brain waves has yet to be born and it appears that it will be some time before one is conceived.

This state of affairs has not inhibited the evolution of a catalog of clinical applications of the ERP. However, this catalog is not as extensive nor comprehensive as that for the EEG. Let's look at some of these applications by considering a new research area known as human–computer interaction. A common analogy today is to relate the computer to the brain just as years ago it was modeled as the mechanism of a clock and 30 or so years ago as a telephone network. We understand how computers work because we have designed them and built the components used in their construction. It is natural that we should try to compare the brain with the computer.

10.5.1   BRAIN'S MAGNETIC FIELDS [41]

The magnetic fields of the brain are also generated by neurons. The electrical signals created in the brain also produce magnetic fields, but these fields are very small. Conditions can be established that lead to the magnetic fields of numerous neurons being summed to produce a signal large enough to be measured at the scalp. In such cases, the neurons must be both large and asymmetric.   The larger the cell, the more capable the cell is of conducting intracellular currents over longer distances, and thus, producing larger magnetic fields.   Asymmetric cells produce magnetic fields that are non–self–cancelling. The pyramidal cells are presently thought to contribute most heavily to the brain's recordable magnetic fields.   Another factor that probably contributes to the summation of magnetic fields is that the contributing cells also meet this criterion.   Further, the cells should also function in unison as they might in event–related potentials or in epileptic discharges.

The flux density of the brain's magnetic field is about two orders of magnitude less than the fields produced by cardiac muscle and eight orders of magnitude less than the earth's magnetic field.   This poses special problems that the recording technology has been addressing.

10.5.2   RESEARCH PROBLEMS

Biomedical signal processing plays a role in monitoring and interpreting the brain's signals produced at the neuronal level and at the more macular level, such as the EEG.   A major problem is to develop inverse filtering procedures that can predict the signals that pass through the skull and skin from the signals that we monitor with electrodes.   The techniques should be applicable to multielectrode arrays, both implanted and on the surface. Improved neural models are needed to help us better interpret the signals we monitor.   These comments apply to all body surface potentials, including the EEG and ECG.   As we learn more about these signals we may be able to develop prosthetic devices for the disabled.   Let us now move on to another application area.

10.6   THE HUMAN–MACHINE CONNECTION

A major effort in computer engineering today is to build machines that execute instructions at an ever faster pace.   The devices to do this contrast with the characteristics of living cells.   Engineering "cells" or building blocks always give the same output for the same input; some devices have memory, some not, but we know which; we have good models of the input/output characteristics; often we do not understand how the devices work at the atomic or molecular level.   We can assemble these devices on chips in various ways to perform numerous functions and for computers these devices do not depend upon the shape of the input or excitation waveform — only on its size.

We can look forward to computers that run faster and faster within the near future.   But for 30 or more years there has been a major bottleneck in computing, namely the input/output (I/O) channel.   This channel has been quite slow because human–machine interfaces must be slow to accommodate the human.   For years, we have had printers, terminals, and keyboards.   These are adequate but awkward.   More recently, the more natural human function of pointing or touching has been accommodated with the advent of touch terminals.

A more advanced interface is the head mounted stereo display that senses the orientation of the viewer's head and displays a simulated graphic environment to the viewer.   There are natural language interfaces and speech recognition and speech synthesis devices. Even eye movements are now being investigated as a means of moving displays. In our laboratory we are exploring the use of monitoring eye movements to allow the user to scan or pan an object or image displayed on a terminal screen. Other human—machine interfaces include software aids such as menu selection of options. Commercial personal computers provide the mouse and joy stick. And some feel that adaptive procedures that adjust to the skill of the human operator will soon be available, e.g., the computer could monitor your typing speed on the keyboard and adjust its response to your skill. Systems are available that can monitor eye movement, pupil size, and thus eye fixation. These systems can be calibrated and can accommodate slight head movements on the part of the person using the system. A TV camera is used to monitor the position of the pupil. The eye movements are mapped onto a pre—calibrated screen which may also contain the scene the user is searching to locate a desired object. Such systems can be used to pan or scroll a screen on a graphics terminal. Pointing is also possible. A more exotic human—machine interface is the direct coupling of the human electroencephalogram (EEG) or event—related potentials (ERPs) (brain waves) to the computer. The EEG or ERP is an indicator of information processing by the brain and the waveform monitored at the scalp is usually discussed in terms of its "components" or positive and negative deflections.

We illustrate this technique with an example. Present a red light to one eye and simultaneously a green light to the other eye. We will see yellow and we will monitor a specific ERP waveform. If we present a yellow light to both eyes, then we will see yellow and get the same waveform as we did before. This ERP response is stable over time, even for years. But it is subject—dependent, meaning each subject has a different waveform.

A more sophisticated experiment is to present subjects with more complicated stimuli, e.g., we may ask subjects to read sentences presented on a computer terminal, controlling for eye movement artifacts, etc. The subject is asked to make a binary decision about these statements, e.g., yes/no or true/false. The statement might be "My name is Don" (see Figure 2). By monitoring several EEG channels, the data is digitized and processed using special algorithms. The output of these algorithms is a prediction of the subject's decision. The task can be performed on—line with typical probabilities of error at 25 percent, with some as low as 2 or 3 percent, depending on the subject. This same method can be used to monitor an individual's level of sensory overload or mental work load. Conceivably, the process can be used to recognize a word that is simply thought by the person [9,10].

A variety of cognitive processes have been associated with particlar ERP components, including anticipation of salient stimuli or responses, attention to a subset of incoming stimuli, and processing of unexpected, rare, or novel stimuli. The meaning of a stimulus has been associated with the relevance of the stimulus for the task — rather than with the semantic or linguistic information stored in long—term memory.

The ERPs just described are elicited by exogeneous stimulus events, i.e., by the sensory or external stimuli. But you can also have ERPs to

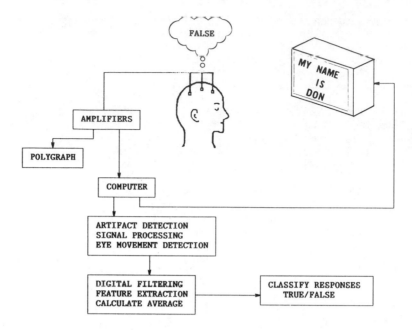

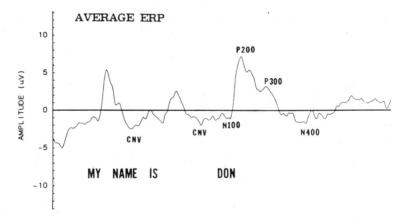

Figure 2   Signal processing for event–related potentials.

endogenous cognitive processes, i.e., internal stimuli brought about by simply thinking or actions of the autonomic nervous system. In other words, you can think a word and have your thought recognized.

One of the most perplexing research problems we face in making this human–machine interface viable is subject variability. While the data for one subject are stable, the features you select for signal processing that subject may not be the same features selected for another subject. This may be partially due to the fact that we are dealing with living biological systems (cells and ensembles of cells) that generate information endogenously. They are not passive — they do not respond merely to an external stimulus as the engineering devices mentioned earlier do. Biological systems can alter their response to an external stimulus. Do these concepts have implications for artificial intelligence? Can we have AI with a computer made of only passive components?

The most frequently discussed human–machine interface is speech. There is talk of a computerized dictating machine. But I am skeptical of the acceptance of such a machine. We have such machines now, in effect. We may dictate and have our dictation transcribed by our secretary. The transcription is not immediate but this service has been available for many years. But yet many do not use this service. Why should they change and use a computerized device? But there are many applications for speech analysis and synthesis techniques. The speech signal carries three types of information:

- What was said
- Who said it
- Information about the environment.

Our interest has been speech synthesis, voice mimicry, and voice creation, as Mel Blanc created the cartoon voices of Bugs Bunny, Porky Pig, and others. What we learn can be used to improve the reliability of speech recognition devices [11].

A common problem in speech recognition is making such systems invariant to the speaker, i.e., speaker independent. Put another way, how can one build a device that converts the speech of all speakers into a canonical speech output? How can we remove the speaker's characteristics from the speech signal? At the present time, we have difficulty extracting subtle features about the speaker from the acoustic or speech signal. We can measure pitch. But other characteristics are more difficult to measure, e.g., the number of formants and their frequency distribution, bandwidth, the shape of the excitation waveform, and other characteristics.

The model we choose to describe our system or our data influences our evaluation of the data. There is one popular speech synthesis model that produces very good speech but which has almost nothing to do with the way humans produce speech, namely, linear prediction. This is analogous to airplanes, they fly but not as birds do. So a computer model may give good results but be irrelevant to answering your questions.

Another problem — perhaps the most important — is just to decide which model to use to analyze your data or how to model the generation of the data. One example is our work to create new voices. We do not yet know how to do this. We can analyze a particular voice and re–create it. But how do we create a new one? Further, there is controversy about how we should even study this problem. We have selected a model of converting one person's speech to another. For example, converting a male voice to a female voice [12] using this model we have found that several factors are important:

1) pitch
2) distribution of the formants
3) bandwidth of formants
4) source–tract interaction, and
5) shape of excitation waveform.

Let us now consider some of the quantitative, analytic signal processing methods used in biomedical signal analysis.

10.7   AVERAGING

One of the most important problems in ERP research is signal extraction; i.e., the extraction of the potential evoked by the sensory stimulus from the ongoing, seemingly random background electroencephalogram (EEG) or noise. If a deterministic, repetitive signal is added to random noise and the time of occurrence of the signal is known, averaging becomes an effective tool for enhancement of the signal [13–16].   Assume that the measured waveform f(t) contains two components: the signal or evoked potential, s(t) and the noise or ongoing EEG, n(t).   If we now assume that the noise is additive, then one can formulate a model of the type,

$$f(t) \quad = \quad s(t) + n(t) \tag{1}$$

If the noise n(t) is independent of the evoked potential and s(t) repeats with each stimulus presentation, then, by simple averaging, one may obtain an estimate of the ERP.

Before the advent of digital computers, Dawson used photographic superposition to obtain a qualitative estimate of the ERP [17].   An oscilloscope was triggered in synchronism with the presentation of a stimulus and a photographic plate was repeatedly exposed to the occurring waveform.   This gave a qualitative and visual estimate of the ERP signal.

Mathematically, the signal–averaging process can be expressed as follows.   Let f(t) be the EEG.   At certain discrete times, a stimulus is presented to the subject.   The signal is then digitized and a single stimulus presentation may be described by the following equation:

$$f_i(t) \quad = \quad \sum_{m=1}^{M} f_i(t_m) \, [u(t-t_m) - u(t-t_m-t_d)] \tag{2}$$

where $f_i(t)$ is the sampled waveform following the presentation of the ith stimulus; u(t) is the unit step function; t_m is the sampling time; $f(t_m)$ is the value of the monitored waveform at the sampling time; and M is the number of samples per stimulus presentation [14].   The sampling duration, t_d is usually very short; its effects may be ignored.   The averaging is then represented by the summation of N waveforms of the same type as equation 2.

$$f(t) = \frac{1}{N} \sum_{i=1}^{N} f_i(t)$$

$$= \frac{1}{N} \sum_{i=1}^{N} \sum_{m=1}^{M} f_i(t_m)[u(t-t_m) - u(t-t_m-t_d)] \tag{3}$$

at a time t_m, the average value becomes

$$f(t_m) = \frac{1}{N} \sum_{i=1}^{N} f_i(t_m) \tag{4}$$

The previous assumption was that the monitored signal $f(t)$ was composed of a signal term $s(t)$ plus a noise term $n(t)$. Thus,

$$f(t_m) = \frac{1}{N} \sum_{i=1}^{N} [s_i(t_m) + n_i(t_m)]$$

$$= \frac{1}{N} \sum_{i=1}^{N} s_i(t_m) + \frac{1}{N} \sum_{i=1}^{N} n_i(t_m) \tag{5}$$

and

$$f(t_m) = s(t_m) + n(t_m) \tag{6}$$

Thus, the average waveform $f(t_m)$ is nothing more than the average signal plus the average noise. Note that if the assumption is made that the signal is deterministic, then

$$s(t_m) = s_i(t_m) \quad \text{for all } i$$

From equation 6, the S/N ratio improves significantly if the noise terms $n_i(t_m)$ are not time locked to the signal. This is usually a valid assumption; thus, the signal may be extracted from the noise by increasing N, the number of replications of the experiment. One can determine the performance of the summation technique by calculating the power signal–to–noise ratio.

The average power (P) in any time–locked signal $s(t)$ is given by

$$P = \frac{1}{m} \sum_{j=1}^{m} s^2(t_i) \tag{7}$$

The average power in the sum of N such potentials is given by

$$\frac{1}{m} \sum_{j=1}^{m} \left[\sum_{i=1}^{N} s_i(t_i) \right]^2 = \frac{1}{m} \sum_{j=1}^{m} [N^2 s^2(t_i)] \tag{8}$$

This is simplified to yield

$$N^2 \cdot \frac{1}{m} \sum_{j=1}^{m} [s^2(t_i)] \ = \ N^2 \cdot P \tag{9}$$

Regarding the noise, we now make the following assumptions:

1.   The expected value of the noise is zero.
2.   The noise in one epoch is uncorrelated with the noise in any other.
3.   The average power is the same for each epoch of noise.

With these assumptions, we now calculate the average noise power in the sum of N potentials.   This quantity is given by

$$\frac{1}{m} \sum_{j=1}^{m} \left[\sum_{i=1}^{m} n(t_i) \right]^2 \ = \ \frac{1}{m} \sum_{j=1}^{m} \left[\sum_{i=1}^{N} [n_i(t_i)]^2 \right] \tag{10}$$

and this can be simplified as:

$$\sum_{i=1}^{N} \frac{1}{m} \sum_{j=1}^{m} [n_i(t_i)]^2 \ = \ NP \tag{11}$$

Thus, the addition of N single evoked potentials will cause the signal power to increase by a factor of N^2, while the noise power will increase by N.

One can also calculate the standard deviation of the average (SDA) with similar results.   First, we compute the expected value of an estimator, $\hat{s}(t)$, of the signal $s(t)$:

$$E[\hat{s}_i(t)] \ = \ E\left[\frac{1}{N} \sum_{i=1}^{N} f_i(t) \right] \ = \ E\left[\frac{1}{N} \sum_{i=1}^{N} (s_i(t)+n_i(t)) \right] \tag{12}$$

$$= \ E\left[\frac{1}{N} \sum_{i=1}^{N} n_i(t) \right] + \frac{1}{N} \sum_{i=1}^{N} E[n_i(t)] \tag{13}$$

Since $s_i(t) = s(t)$ for all i (assumption of signal invariance) and $E[n_i(t)]=0$ for all i, we have

$$E[\hat{s}(t)] \ = \ s(t) \tag{14}$$

Now we compute

$$(SDA)^2 = E[\hat{s}(t) - s(t)]^2 = E\left[\frac{1}{N}\sum_{i=1}^{n} n_i(t)\right]^2 \qquad (15)$$

$$= \frac{1}{N^2}\sum_{i=1}^{N}\sum_{j=1}^{N} E[n_i(t)n_j(t)] = \frac{1}{N^2}\sum_{i=1}^{N} E[n_i^2(t)] \qquad (16)$$

Equation 16 holds as long as $E[n_i(t)n_j(t)] = 0$ for all $i \neq j$ (i.e., noise samples are uncorrelated). Therefore,

$$(SDA)^2 = \frac{1}{N^2} \cdot N\sigma^2 \qquad (17)$$

where $\sigma^2 = E[n_i^2(t)]$ and

$$SDA = \frac{\sigma}{\sqrt{N}} \qquad (18)$$

Thus averaging decreases the noise in a manner directly proportional to the standard deviation of the noise and in a manner inversely proportional to the square root of the number of replications.

The pitfalls of conventional averaging are primarily caused by invalid assumptions made in the model. Let us briefly review these. First, the monitored scalp brain wave $f(t)$ is made up of a signal component, $s(t)$, plus a noise component, $n(t)$, i.e.,

$$f(t) = s(t) + n(t) . \qquad (19)$$

This assumption implies that the noise is purely additive.

Second, the signal is purely deterministic, i.e.,

$$s_i(t) = s_j(t) \quad \text{for all } i \text{ and } j . \qquad (20)$$

This assumption implies that for the same stimulus (and subject), the signal will faithfully repeat itself with every stimulus replication.

Third, the noise is not time–locked to the presentation of the stimulus. This assumption implies that $E[n_i(t)] \to 0$ as i increases.

Of these assumptions, the second is probably most difficult to accept. The shifting of signals in time with stimulus replication is a generally accepted premise. The ambiguity present in average response records was recognized as long as 20 years ago.

10.7.1   CROSS–CORRELATION AVERAGING [14]

As was mentioned earlier, the signal, or single ERP, probably varies on a stimulus–to–stimulus basis. One of the earliest attempts to take this variability into account was proposed by Woody. His iterative filter algorithm attempts to compensate for the random fluctuations in latency displayed in the entire signal. The approach is to cross–correlate an essentially arbitrary template with a sequence of samples of the response in order to estimate the signal latency. The individual signals are then corrected for their individual latency variations and a new average signal is computed. Often, the conventional average is chosen as the initial template. This process may be repeated with each new average signal used as the template; further refinements in the latency estimates are obtained and new average signals are computed.

10.7.2   MEDIAN AVERAGING [14]

Typical ERP studies use 100 repetitions of the same stimulus to produce an average ERP or signal. Frequently, estimates of the signal after 10 or 20 repetitions are desired.

Computation of the median evoked response is conceptually very simple. At each latency, the amplitude values are arranged in ascending order of magnitude, then the median at the particular latency M_e is given by the $(n+1)/2$ value. If n (the number of values) is odd, the median is the middle value of the set of ordered data; if n is even, the median is usually taken as the mean of the two middle values of the set. The use of the median becomes important when the number of sample functions is contaminated by noise, the median gives a less biased estimate of the signal. For a large number of samples, the mean is a better estimate than the median.

10.7.3   LATENCY–CORRECTED AVERAGING [14]

There is evidence that indicates that individual peaks in each successive single trial ERP may independently shift from response to response. The technique of latency–corrected averaging (LCA) is aimed at the quantification and correction of the latency jitter of the different components of the single evoked potential.

The processing techniques required to compute the LCA may be divided as follows:

1.   Data preprocessing through a minimum mean square error (MMSE) filter.
2.   Peak identification and selection within the single ERP.
3.   Computation of the histogram of peak latencies.
4.   Selection of groups of peaks having statistical properties different from noise.
5.   Aligning and averaging peaks in the same intervals.
6.   Computation of LCA component statistics.

These will now be briefly discussed, one by one.

(1)   Data Processing

After digitization, the sequence of N data points resulting from the ith stimulus is assumed to be made up of the sum of a signal sequence s_k and a noise sequence n_k.   Once all of the data sequence have been digitized and stored, a minimum mean square error filter (MMSE) is designed to remove as much of the ongoing EEG as possible while only minimally affecting the ERP components.   The design of the filter is accomplished by solving the set of equations

$$R \underline{h} = \underline{g} \tag{27}$$

where

$$R = \begin{bmatrix} r_0 & r_1 & r_2 & \cdots & r_{M-1} \\ r_1 & r_0 & r_1 & \cdots & r_{M-2} \\ \vdots & & & & \vdots \\ r_{M-1} & & \cdots & & r_0 \end{bmatrix} \tag{28}$$

$$\underline{h} = \begin{bmatrix} h_0 \\ h_1 \\ \vdots \\ h_{M-1} \end{bmatrix} \qquad \underline{g} = \begin{bmatrix} g_0 \\ g_1 \\ \vdots \\ g_{M-1} \end{bmatrix} \tag{29}$$

The first row of R is the sample autocorrelation of the measured data. The column vector $\underline{h}$ consists of the filter coefficients, and the column vector $\underline{g}$ is the sample cross–correlation vector between the true signal and the measured data.   A first estimate of the signal is considered to be equal to the average data vector.   The set of equations are rather easily solved since R is a Toeplitz matrix and is easily inverted.   The solution of Equation 27 is given by

$$\underline{h} = R^{-1} \underline{g} \tag{30}$$

Once $\underline{h}$ is found, each data vector is convolved with $\underline{h}$ giving the MMSE filtered version of the data set.   (See Paper 4 by Van Loan for a discussion of matrix formulated problems and their solutions).

(2)   Peak Identification and Selection within the Single ERP

Following the filtering process outlined above, each filtered waveform is searched with a prescribed template and the output designated as a peak when certain predetermined criteria are met.

(3)   Computation of the Histogram of the Peak Latencies

Following peak identification a histogram is constructed of the number of peaks detected at each latency.

(4)   Selection of Groups of Peaks having Statistical Properties Different From Noise

A running mean over a specified interval of the histogram is constructed. This gives the running mean. The zero crossings and positive and negative epochs of the signal will be evident. In order to provide a systematic procedure for determining which peaks should be grouped together, the running mean at each point is tested at a specified confidence level using the nonparametric sign test against the hypothesis that the samples were drawn from a zero–mean population. This procedure partitions the latency interval into regions in which components having the same sign are consistently found, and at the same time, specifies which peaks should be assigned to each particular interval.

(5)   Aligning the Peaks for Averaging in the Same Intervals

For each component interval, all of the responses for a given subject which are detected in this interval are aligned so that the peaks coincide. These latency–corrected waveforms are then averaged over a specified range in the vicinity of the peak, and the resulting waveform is reproduced at the mean latency for this particular component. This is repeated for all components giving an overall latency–corrected average for each subject.

(6)   Computation of the Statistics of the Components of the LCA

Once the peaks have been detected and assigned to the appropriate latency intervals, the means and standard deviations for the responses in each interval are computed.

The above technique produces statistics for each of the components detected and a waveform which is considered to be an improved estimate of the signal or single ERP. An application of this technique is separation of single trials into homogeneous sets by finding those trials which contain specific combinations of peaks.

10.7.4 SUMMARY AND CONCLUSIONS

Although conventional averaging remains one of the most used signal processing tools in evoked potential research, it must be used with care and an understanding of its limitations. Other techniques have recently surfaced which attempt to correct some of the inherent deficiencies present in conventional averaging. Cross–correlation averaging must be well understood before an experimenter chooses any of these methods to improve the information content of an "averaged" waveform.

Signal averaging is a form of comb filtering. This can be shown by calculating the transfer function of the signal averaging computer.

For the noise–free case the signal is observed through an observation window, $w(nT)$ of duration T_0. The average is then given by

$$f(nT) = \frac{1}{M} \sum_{k=0}^{M-1} s(nT-kT_0)w(nT-kT_0) \tag{31}$$

$$= \frac{1}{M} \sum_{k=0}^{M-1} s(nT-k(T_0/T)T)w(nT-k(T_0/T)T) \tag{32}$$

where N is the number of samples in the observation time T_0 and M is the number of signal repetitions. The z–transform of Equation (32) is

$$F(z) = \sum_{n=0}^{N-1} \left\{ \frac{1}{M} \sum_{k=0}^{M-1} s(nT-k(T_0/T)T)w(nT-k(T_0/T)T) \right\} z^{-n} \tag{33}$$

which can be simplified if w(nT) is the rectangular window, i.e., w(nT)=1. Then interchanging the order of summations in Equation (33), we get

$$F(z) = \frac{1}{M} \sum_{k=0}^{M-1} S(z) \, z^{-k(T_0/T)}$$

$$= S(z) \frac{1}{M} \sum_{k=0}^{M-1} z^{-k(T_0/T)}$$

$$= S(z) \left[\frac{1}{M} \frac{1 - z^{-M(T_0/T)}}{1 - z^{-(T_0/T)}} \right] \tag{34}$$

Therefore, the averaging process can be represented as a filtering of the basic signal waveform where the filter transfer function is given by

$$H(z) = \frac{1}{M} \frac{1 - z^{-M(T_0/T)}}{1 - z^{-(T_0/T)}} \quad , \tag{35}$$

the discrete frequency response of which is found by letting $z = e^{j(2\pi/N)m}$:

$$H(mf) = \frac{1}{M} \frac{1 - e^{-j2\pi mM \frac{T_0}{NT}}}{1 - e^{-j2\pi m \frac{T_0}{NT}}}$$

$$= \left[e^{-j2\pi m \frac{T_0}{2NT} (M-1)} \right] \left[\frac{1}{M} \frac{\sin \frac{\pi m M T_0}{NT}}{\sin \frac{\pi m T_0}{NT}} \right] \tag{36}$$

which is recognized as a form of comb filter. The magnitude squared of H(mf) may be plotted for various values of M, the number of averages performed. As M increases, the comb filtering action of the computer averaging process becomes more apparent, i.e., the filter becomes sharper about multiples of $1/T_0$.

The above discussion is limited to the case where the signals are periodic, although the period of the signal may not necessarily be restricted to be the same as the observation (window) interval.

10.8   FILTERING [15,19,20]

Evoked brain potentials are measured with the aid of many scalp electrodes. These measurements are referenced to a body position, such as linked ears, that is some distance from the location on the cortex where the response is expected.   The amplitudes of the ERPs vary from tenths of a microvolt to a few microvolts (μV), and are embedded in the ongoing EEG.   The signal—to—noise ratio (SNR) is less than 1:1 (0 dB).   In some cases the ratio of the signal voltage to EEG voltage may be as small as 1:10 giving a SNR of —20dB.   It is this small SNR that makes waveform estimation and signal classification difficult.

The key to successful information extraction is the development of a mathematical model suitable for representation of the ERP waveform and then employing this model to design appropriate processors.   Various models have been proposed, but the one that is most commonly used and which has given the most consistent results is the one in which the waveform measured at the scalp is assumed to be signal plus noise:

$$r(t) \quad = \quad s(t) + n(t) \tag{37}$$

where $r(t)$ is the measured waveform, $s(t)$ is the component of the measured waveform associated with the event—related (evoked) potential, and $n(t)$ is the component of the measured waveform independent of the evoked potential.   In most practical cases, $n(t)$ is taken to be the ongoing EEG.

The signal $s(t)$ in Equation 37 can be extracted from $r(t)$ by a variety of signal processing techniques, the details of which depend on the measured or assumed characteristics of both $s(t)$ and $n(t)$.   The most widely reported techniques are based on an assumption that the noise and often (with dubious validity) the signal are sample functions from stationary ramdom processes. Such an assumption leads to Wiener filters or Kalman filters.   Other assumptions lead to filters of different types such as the minimum mean square error (MMSE) or maximum signal—to—noise ratio (MSNR) filters.   If a priori information about the waveshape is available it can often be incorporated into the signal processor to further improve performance.   Significant improvements in waveform estimation have been achieved with filters of these types, but further improvements are required before a fully useful level of performance is attained.

The bandwidth of ERP's extends from below 1 Hz, perhaps to nearly DC, to about 1000 Hz.   But the high frequency components are generally the early time domain components and these may often be filtered out.   Typical filter bandwidths range from 0.1 Hz (or lower) to 50 Hz (or perhaps 100 Hz).

Most ERP researchers are concerned with the analysis of ERP components in the time domain. Frequently, the task is to relate a component to a cognitive task, such as decision making.

10.8.1   WIENER FILTERS

Probably the filter most widely applied to ERP processing outside the low–pass filter is the Weiner filter. This filter was originally developed for signal estimation in communication and control applications. As generally implemented, the underlying assumptions for this filter to be optimum (in the sense of minimizing the mean squared error of estimate) are that the signal and noise be additive, statistically independent, and that each be a sample function from a wide–sense stationary random process. There is some question as to whether these assumptions are met by evoked potential waveforms both with regard to the signal (ERP) being a stationary random process and the noise and signal being statistically independent. Nevertheless, many researchers have used filters of this type and have arrived, perhaps not too unexpectedly, at contradictory conclusions as to their effectiveness. Under the assumption given above, the transfer function for the Wiener filter for a signal embedded in additive noise can be written in the frequency domain as

$$H_1(\omega) \quad = \quad \frac{S_s(\omega)}{S_s(\omega) \, + \, S_n(\omega)} \tag{38}$$

where $S_s(\omega)$ is the spectral density of the signal and $S_n(\omega)$ is the spectral density of the noise. For processing the average event related potential, the Wiener filter becomes

$$H_2(\omega) \quad = \quad \frac{S_s(\omega)}{S_s(\omega) \, + \, \frac{1}{K} \, S_n(\omega)} \tag{39}$$

where K is the number of responses used to obtain the average. These formulations are noncasual in that their inverse transforms (i.e., the filter impulse responses) are nonzero for negative time. This is not a serious problem but does require that the entire waveform be available for processing every point. Also, there are likely to be edge effects when this filter function is used to process truncated segments of a waveform. This effect can be minimized by applying a window function to the data causing it to decrease smoothly to zero at the edges. The problem can also be handled by processing the data in the time domain with a filter function that has a finite duration. If no care is given to this problem and the data are processed using the FFT it is likely that distortion will be introduced by the processing itself.

There are a number of subtleties involved in the application of Wiener filters to empirically derived data. A major source of difficulty is the necessity of estimating the spectral densities of the (ERP) and the noise (ongoing EEG). Frequently these estimates are determined by averaging the magnitude squared of the spectra of the individual responses. Let the measured waveform for the ith stimulus be

$$x_i(t) = s(t) + n_i(t) \tag{40}$$

and let the ensemble average $\bar{x}(t)$ be

$$\bar{x}(t) = \frac{1}{N} \sum_{i=1}^{N} x_i(t) = s(t) + \frac{1}{N} \sum_{i=1}^{N} n_i(t) \tag{41}$$

For the case of uncorrelated signal and noise it is readily shown that

$$S_{\bar{x}}(\omega) = S_s(\omega) + \frac{1}{N} S_n(\omega) \tag{42}$$

$$S_x(\omega) = S_s(\omega) + S_n(\omega) \tag{43}$$

where $S_{\bar{x}}(\omega)$, $S_s(\omega)$, $S_n(\omega)$ and $S_x(\omega)$ are the spectral densities of the ensemble average of $x(t)$, the signal $s(t)$, the noise $n(t)$, and the random process $x(t)$, respectively.

Solving these equations for $S_s(\omega)$ and $S_n(\omega)$ gives

$$S_s(\omega) = \frac{N}{N-1} \left[S_{\bar{x}}(\omega) - S_x(\omega) \right] \tag{44}$$

$$S_n(\omega) = \frac{N}{N-1} \left[S_x(\omega) - S_{\bar{x}}(\omega) \right] \tag{45}$$

The Wiener filter transfer function for the a posteriori processor is then

$$H_2(\omega) = \frac{S_s(\omega)}{S_s(\omega) + \frac{1}{N} S_n(\omega)} = \frac{N}{N-1} \left[1 - \frac{1}{N} \frac{S_x(\omega)}{S_{\bar{x}}(\omega)} \right] \tag{46}$$

A major difficulty arises in determining the proper values of $S_x(\omega)$ and $S_{\bar{x}}(\omega)$ as this falls into the category of spectral estimation. A number of investigations use the squared magnitude of the Fourier transform to estimate the spectral density. Thus,

$$S_x(\omega) = \frac{1}{N} \sum_{i=1}^{N} |\{X_i(t)\}|^2 \tag{47}$$

$$S_{\bar{x}}(\omega) = \left| \frac{1}{N} \sum_{i=1}^{N} \{X_i(t)\} \right|^2 \tag{48}$$

Generally speaking this is not a desirable way to estimate power spectra as such estimates are biased and tend to have unnecessarily large variances. More satisfactory procedures are to estimate the spectral density as the Fourier transform of the autocorrelation function after appropriate windowing. Alternatively, the spectral density can be estimated from the smoothed periodogram, which is equivalent to convolving the functions of Equations 47 and 48 with an appropriate window function. We discuss spectral estimation in greater detail in another section. There have been favourable and unfavorable results reported with regard to the use of Wiener filters [15].

10.8.2   OTHER FILTER DESIGNS

One of the important quantities that is measured in ERP research is the latency of components, and because of this it is desirable to preserve the time relations of waveforms when filtering is performed. The time delay through a filter is proportional to the derivative of the phase with respect to frequency. By making this phase shift constant or zero, the time relationships will be retained. This is readily accomplished in the time domain by designing filters that have symmetry about the origin. Such filters are inherently noncausal, but by accepting a time delay that is one half the filter duration, the filter becomes casual. Methods for designing linear phase are now readily available.

A special filter for use in processing evoked potentials is described in [15]. The filter provides a minimum mean square error estimate of the signal from samples of the signal plus independent noise. The filter function expressed as a weighting vector in the time domain is obtained as the solution of the matrix equation.

$$\underline{h} = R^{-1} \underline{g}$$

where $\underline{h}$ is the filter vector, R is the sample correlation matrix, and $\underline{g}$ is the sample signal crosscorrelation vector. Thus

$$r_{j-k} = \frac{1}{M + N - 1} \sum_{t=0}^{M+N-2} d_{t-k} d_{t-j} \tag{49}$$

$$g_j = \frac{1}{M + N - 1} \sum_{t=0}^{M+N-2} s_t d_{t-j} \tag{50}$$

where d_j is the jth element of the data vector, and s_j is the jth sample of the signal vector. It is assumed that there are N data samples and M filter coefficients. The sample correlation matrix R is a Toeplitz matrix for which simple inversion techniques are available. The principal problem in using this procedure is to estimate the elements R and g.

A similar processor is obtained if the expected value of the MSE is minimized instead of the error itself. In such a case R is the ensemble autocorrelation matrix and the vector g is the ensemble crosscorrelation vector between the signal and the data set. This procedure requires that the signal and noise be wide–sense stationary processes, which is difficult to justify because of the assumed deterministic nature of the signal. When stationarity is not assumed the values of R and $\bar{g}$ are obtained by averaging over the data set. It is shown that R is given by

$$R = \begin{bmatrix} r_0 & r_1 & r_2 & \cdots & r_{M-1} \\ r_1 & r_0 & r_1 & & r_{M-2} \\ \vdots & \vdots & & & \vdots \\ r_{M-1} & r_{M-1} & \cdots & & r_0 \end{bmatrix} \qquad (51)$$

where the vector $r = (r_0, r_1, ..., r_{M-1})^t$ has components that are the averages of the individual autocorrelation vectors for each segment. The components of $\bar{g}$ are given by

$$g_j = \frac{1}{M+N-1} \sum_{t=0}^{M+N-2} \bar{s}_t \bar{s}_{t-j} \qquad (52)$$

where $\bar{s}_t$ is the tth element of the mean of the data vectors. This filter has been used but needs to be compared with other filters [15].

10.9   SPECTRAL ANALYSIS [16,21–25]

Spectral analysis has been used for exploring the structure of a data set. Scientific investigations usually seek to find features in the spectrum that are correlated with physiological events or with disorders.

10.9.1   AUTOCORRELATION METHOD AND PERIODOGRAM

There are two equivalent definitions for the power spectrum. If we denote

$$X_N(z) = \sum_0^{N-1} x_n z^{-n} \qquad (53)$$

Then the power spectrum is

$$S(\omega) = \lim_{N \to \infty} \frac{1}{N} E \left| X_N(z) \right|^2_{z=e^{j\omega t}} \qquad (54)$$

where E denotes the expected value. This is the direct definition of the power spectrum since it is based on the z–transform of the data evaluated on the unit circle. This is rather easily shown to be equivalent to the z–transform of the autocorrelation function.

$$S(\omega) = \lim_{N\to\infty} \sum_{-(N-1)}^{N-1} (1- \frac{|m|}{N})R_m z^m \tag{55}$$

where the autocorrelation function is defined for stationary random processes as

$$R_m = R(mT) = E(x_n x_{n-m}) = E(x_n x_{n+m}) \tag{56}$$

This is known as the autocorrelation definition of the spectrum since it is based on the z–transform of the autocorrelation function evaluated on the unit circle.

In practice, the ensemble average of the data is replaced with an estimation procedure such as time averaging. One commonly used estimate is the periodgram.

$$\hat{S}(\omega) = \frac{1}{N} |x_N(\omega)|^2 \tag{57}$$

The quality of the spectral estimate is then evaluated by determining whether the estimate is biased or not and by determining if the variance and the bias of the estimate decreases as the data record length (number of data samples, N, for a fixed sampling interval, T) increases. A consistent estimate is one which possesses this latter property. The bias of an estimator is defined as the difference between the expected value of the estimate and the true value of the quantity being estimated, which in this case is the true power spectrum, and is therefore not a random variable nor a random process.

Two commonly used estimates for the autocorrelation function are

$$\hat{R}_k = \frac{1}{N-|k|} \sum_{0}^{N-|k|-1} x_n x_{n+|k|} \tag{58}$$

which is unbiased, i.e., $E(\hat{R}_k) = R_k$, and

$$\hat{R}_k = \frac{1}{N} \sum_{0}^{N-|k|-1} x_n x_{n+|k|} \tag{59}$$

which is biased since

$$E[\hat{R}_k] = \frac{N-|k|}{N} R_k \ . \tag{60}$$

Either of these estimates may be used to yield an estimate of the spectrum.

The variance of these estimates is discussed in most modern text books on signal processing. The estimate selected for the autocorrelation function is usually the one with the minimum bias and minimum variance relative to the true autocorrelation function. Such estimates are said to have good statistical stability and yield good power spectral density estimates.

The statistical stability of the spectral estimate may also be increased by windowing the estimate of the autocorrelation function before transforming to the frequency domain. Windowing will generally reduce the variance of the power spectral estimate. But the various window functions are generally not data–dependent and provide predictable results regardless of the data being analyzed.

Two factors affecting the frequency resolution of the spectral estimate are the finite record length of the autocorrelation function (and the data) and the window applied to the autocorrelation function. As the record length increases the spectral resolution usually increases (but not always, e.g., the periodogram of white noise just becomes more oscillatory). And if the window is designed to reduce the leakage (or sidelobes) in the spectral domain, then the spectral resolution is decreased.

Using the above conventional methods, increased resolution is achieved by appending a sequence of zeros to the windowed autocorrelation functions prior to transforming to the frequency domain. But this only decreases the spacing between spectral line components in the discrete Fourier transform and does not really improve the resolution between two closely spaced spectral components of the signal. The more modern techniques of autoregressive (AR), linear predictive (LP), and maximum entropy (ME) spectral analysis achieve increased spectral resolution, particularly for short data records. These equivalent methods in effect extend the autocorrelation function by extrapolating the autocorrelation function rather than appending zeros.

Besides windowing, the variance of the power spectral density may be reduced by segmental averaging. This procedure segments the data record and calculates an estimate of the autocorrelation function for each data segment. The average of these autocorrelation estimates is then determined. The estimated power spectrum is the discrete Fourier transform (DFT) of the average estimated autocorrelation function. These power spectral estimates are biased and the bias exceeds the bias of the periodogram of the unsegmented data record. This is because the main lobe of the segmented data is less than the variance of the periodogram by a factor equal to the number of segments if the segments can be considered independent (or uncorrelated) Gaussian segments. But the segmented estimate has less resolution than that of the periodogram of the entire data record.

10.9.2 SPECTRAL ANALYSIS BY DATA MODELS

During the 1960's two nonlinear spectral estimation techniques were introduced. One is known as the maximum likelihood method (MLM). The other is known by at least three names, e.g., the autoregression (AR), the linear prediction (LP), or maximum entropy method (MEM). From the point of view of spectral analysis these methods are particularly attractive for making high resolution spectral estimates when the data record is short. Neither method uses window functions and the methods are considered data–adaptive. For ERP and EEG analysis, these techniques, may not offer significant advantages over the conventional methods in all cases. This is because the data records for ERP and EEG are usually long and the sampling rate is relatively high with respect to the highest frequency in the data. This means that the resolution in the frequency domain is usually quite adequate.

However, there are occasions when high resolution spectral analysis is needed and the data records are short. Also, there are times when one may wish to predict or extrapolate the data or the autocorrelation funciton. In these situations these methods become very useful.

Perhaps, the major advantage these new methods offer is that they provide a mechanism for modeling the data. This requires the researcher to think about the limitations of the data model and to seek ways to improve it and therefore obtain better analysis results.

The principle behind the AR method is that increased spectral resolution can be obtained by predicting or extrapolating the autocorrelation function beyond its data limited range.. If the data available is of limited duration, then the autocorrelation function duration is also limited, and further, the autocorrelation values for large lags (shifts in the data) are unreliable due to the relatively few overlapping data samples. The new procedure provides a method for extending the autocorrelation function when the data is not available. The principle used for this prediction process is that the spectral estimate must be most random (the spectrum will be the flattest or the most white) or have the maximum entropy of any power spectrum which is consistent with the known sample values of the autocorrelation function. The objective is to add no new information. The derivations for this method and others may be found elsewhere [21].

10.9.3   SURVEY OF DATA MODELS

We may represent a sampled time series as either

$$x(nT) \ = \ x(n) \ = \ x_n \tag{61}$$

where T is the sampling interval. The one–sided z–transform of this series is given by

$$X(z) \ = \ \sum_0^\infty x_n z^{-n} \tag{62}$$

The "spectrum" is found by letting $z = e^{j\omega T}$ and we frequently suppress T or assume T = 1. Thus

$$X(e^{j\omega T}) \ = \ X(\omega) \ = \ X(z)\Big|_{z \ = \ e^{j\omega T}} \tag{63}$$

Three basic data models (note these are data models that lead to spectral models) are autoregressive (AR), moving average (MA), and autoregressive–moving average (ARMA).

The autoregressive model produces a data sequence y_n, (or an output) when excited by a white noise sequence, x_n, as follows

$$y_n \ = \ a_1 y_{n-1} + a_2 y_{n-2} + \ ... \ + a_n y_{n-N} + b_0 x_n$$

$$y_n \ = \ \sum_1^N a_i y_{n-i} + b_0 x_n \tag{64}$$

or

$$Y(z) = \frac{b_0}{1 - \sum_{1}^{N} a_i z^{-i}} X(z) \qquad (65)$$

Thus, the autoregressive data model corresponds to an all–pole system model, with N determining the number of past output values used in the recursion relation.

The moving average data model is given by

$$y_n = b_0 x_n + b_1 x_{n-1} + \dots + b_k x_{n-k} = \sum_{0}^{k} b_i x_{n-i} \qquad (66)$$

or

$$Y(z) = \left[\sum_{0}^{k} b_i z^{-i} \right] X(z) \qquad (67)$$

which says that the moving average data model is an all–zero system model, with k determining the number of past input values used in the recursion relation.

Finally, the mixed data model is the autoregressive–moving average data model given by

$$y_n = \sum_{1}^{N} a_i y_{n-i} + \sum_{0}^{k} b_i x_{n-i} \qquad (68)$$

or

$$Y(z) = \frac{\sum_{0}^{k} b_i z^{-i}}{1 - \sum_{1}^{N} a_i z^{-i}} X(z) \qquad (69)$$

which is a pole–zero system model.

A special case is the prediction of y_n using a linear combination of past values of the output, e.g.,

$$\hat{y}_n = \sum_{1}^{N} a_i y_{n-i} \qquad (70)$$

There will, of course, be an error between this predicted value and the actual value, i.e.,

$$e_n = y_n - \hat{y}_n = y_n - \sum_{1}^{N} a_i y_{n-i} \tag{71}$$

or

$$y_n = \hat{y}_n + e_n \tag{72}$$

which may be expressed in the z domain as

$$Y(n) = \cfrac{1}{1 - \sum_{1}^{N} a_i z^{-i}} E(z) \tag{73}$$

This is an AR model, where e_n may be considered the input or driving function or y_n may be the input and e_n the output. In the latter case we attempt to select the a_is so as to minimize the error, e_n. In either case the a_is are usually found by minimizing the mean–square error. These coefficients are known as the autoregressive coefficients and are the same as the linear predictive (LP) coefficients, and they are also known as the maximum entropy (ME) coefficients.

10.9.4 STATISTICAL CONSIDERATIONS

The most frequently used spectral analysis procedure is the periodogram, i.e., simply the magnitude squared of the Fourier transform of a finite record length of data. This is equivalent to the transform of the estimate of the autocorrelation function. But, while the periodogram is an unbiased estimator, it is not consistent, i.e., its variance does not decrease as the data record length increases. In fact, one can show that as the data record length increases the spectrum becomes more "oscillatory" in nature. What can be done to improve our spectral estimates?

First, recall that we evaluate our estimated by two fundamental statistical measures. These are the mean or expected value of the estimate and the variance of the estimate. We usually want the mean of the estimate to be equal to the true quantity being estimated, i.e., we want the estimate to be unbiased; the true spectrum minus the mean of the spectral estimator would be zero. If the estimator is biased, then we would like the bias to decrease as the record length (number of measurements) increases. Similarly, the variance of the estimate should decrease as the record length increases. If these latter two conditions prevail, then the estimate is consistent.

Given that one has an unbiased estimator, then how can one reduce the variance of this estimate? Two major approaches exist, which in fact are essentially equivalent. The first method is to average spectral estimates made from individual time records of the data. These records can be juxtaposed or overlapped, they can be windowed or not. The general outcome is that the variance of the final estimate is less than the estimate for the individual records.

The second approach, which can be shown to be essentially the same as the above segmental averaging procedure, is to window one long data record. This total data record is the same length as the sum of the juxtaposed individual record lengths used for segmental averaging. The amount of the variance reduction obtained is dependent upon the type of window employed. The penalty paid to obtain a more statistically stable spectral estimator is a loss in spectral resolution, i.e., two closely spaced frequency components may become blurred into one spectral peak. These remarks apply to the more conventional spectral estimators. What about the newer methods? To date no good estimate of the variance of the AR (LP or ME) estimator exists. More research is needed.

The advantage of the LP/AR/ME approach is that a formal all–pole model is applied to estimate the spectrum. The resolution of this estimator is greater than previous methods but the estimator can be sensitive to the signal–to–noise ratio of the data. An advantage is that the computational complexity of these newer methods is on the same order of magnitude as that for the more established conventional methods. Finally, the LP method is data–adaptive because of the model, i.e., new coefficients are determined for each data set, and no windows are needed.

10.9.5   OTHER SPECTRAL ANALYSIS METHODS [21–24]

Pisarenko's method was developed to produce a line spectrum for those cases when the data is best modeled as sinusoids in noise. In this case the AR (LP or maximum entropy) method is not very suitable since it is designed to produce the flattest possible spectral estimate. Pisarenko's algorithm is not complicated and the Pisarenko method also gives good resolution.

For high resolution spectral line estimation the eigenvalue decomposition technique should be considered. This method seems to be more robust than Pisarenko's in the presence of colored noise or covariance errors and also seems better for those cases where the computation is limited to finite precision. There are several algorithms to be considered. One approach is to solve for the roots of a polynomial obtained from the covariance matrix of the data. Another approach is to use a state–space representation and perform a singular value decomposition on the estimated covariance matrix. These approaches make extensive use of matrix computations and inversions, which today can be done with numerical stability.

10.10   ADAPTIVE FILTERING

As we have seen, signal averaging and conventional filtering or signal conditioning have been popular approaches to improve the signal–to–noise (SNR) in EEG and ERP research. But another approach is adaptive filtering [26,27]. Other biomedical applications have included removal of 60 Hz power line interference, removal of electrosurgical interference, and cancellation of maternal ECG from fetal ECG.

What is an adaptive filter? Such a filter can adjust its parameters to obtain an optimum performance using a specified error criterion for changing

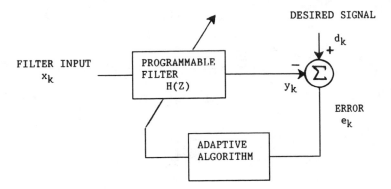

Figure 3   Schematic of adaptive filter.

environmental (input) characteristics.   An adaptive filter architecture is pictured
in Figure 3.   Two processes are present in this filter:

1.   The adaptive or training process, which addresses the
         automatic adjustment of the filter coefficients.

2.   The filtering process, which uses the adaptive filter
         coefficients to produce an output by weighting the input
         signals as with the tapped delay line filter, i.e., (see Fig. 4)

$$y_n = \sum_{k=0}^{p} w_k x_{n-k} \tag{74}$$

Why is an adaptive filter needed?   Fixed filter designs usually require
a priori information of the input signal or perhaps the statistics about these
signals.   In many applications such information is unknown or must be
assumed.   Further, the signals may vary with time.   In these cases the fixed
filter design will be suboptimal.   The adaptive filter, in principle, has the
capability to learn the input signal characteristics and track the signal by using
an adaptive algorithm to produce the output with minimum error.

What are the major concerns when designing an adaptive filter?   There
is the filter's convergence rate.   If the input varies, how fast (number of
iterations required) will the filter move toward its steady state (minimum error
rate state)?   The performance of the filter is also of interest.   When the filter
is in the steady state, what is the error between the desired and the actual
outputs?

Figure 4 gives several possible adaptive filter structures, algorithms
used for their design, and applications.

Let use consider first the Weiner filter, which is similar to the tapped
delay line filter shown in Figure 4.   We have a desired signal and a noisy

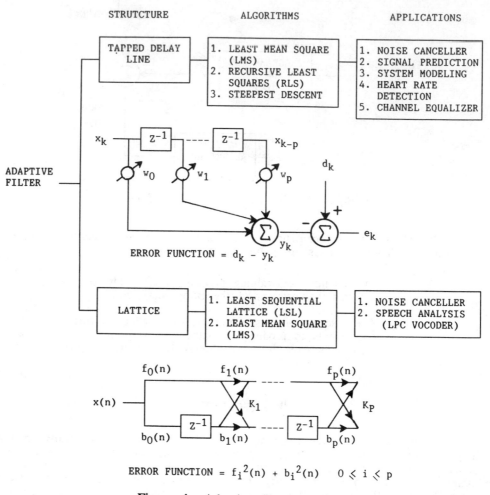

Figure 4 Adaptive filter overview.

signal, which we wish to remove to give an estimate of the desired signal. The estimated signal is given by

$$y_n = \sum_{k=0}^{p} w_k x_{n-k} = (w_0, w_1, \ldots, w_p) \begin{bmatrix} x_n \\ x_{n-1} \\ \vdots \\ x_{n-p} \end{bmatrix}$$

$$= \underline{w}^t \underline{x}(n) \tag{75}$$

The filter weights w_k are optimally selected by minimizing the mean–squared error,

$$E[e_n^2] = \text{minimum} = E[(d_n - y_n)^2] \tag{76}$$

where d_n is the desired signal and E denotes the expected value. This minimization criterion leads to the orthogonality conditions,

$$E[e_n x_{n-m}] = 0 \qquad 0 \le m \le p \tag{77}$$

or

$$E[e_n \underline{x}(n)] = 0 \tag{78}$$

This gives the normal equations

$$E[(d_n - y_n)\underline{x}(n)] = E[(d_n - \underline{w}^t \underline{x}(n))\underline{x}(n)] = 0 \tag{79}$$

or

$$E[\underline{x}(n)\underline{x}^t(n)]\underline{w} = E[d_n \underline{x}(n)] \tag{80}$$

which is

$$R\underline{w} = \underline{r} \tag{81}$$

where

$$R = E[\underline{x}(n)\underline{x}^t(n)] \quad \text{and} \quad \underline{r} = E[d_n \underline{x}(n)] \tag{82}$$

The optimum Weiner filter weights are given by

$$\underline{w} = R^{-1}\underline{r} \tag{83}$$

and the minimum error is

$$\begin{aligned}
E[e_n^2] &= E[e_n(d_n - \underline{w}^t \underline{x}(n))] = E[e_n d_n] \\
&= E[(d_n - \underline{w}^t \underline{x}(n))d_n] \\
&= E[d_n^2] - \underline{w}^t E[\underline{x}(n)d_n] \\
&= E[d_n^2] - \underline{w}^t \underline{r} \\
&= E[d_n^2] - \underline{r}^t R^{-1}\underline{r}
\end{aligned} \tag{84}$$

since R and R^{-1} are symmetric.

The desired signal can be expressed as

$$d_n = e_n + \underline{w}^t \underline{x}(n) \tag{85}$$

The first (error) term is uncorrelated with $\underline{x}(n)$ while the second term is correlated with $\underline{x}(n)$. The filter removes from the desired signal any part that is correlated with the input $\underline{x}(n)$. The remaining part, e_n, is uncorrelated with $\underline{x}(n)$. Thus, the Weiner filter is known as a correlation canceler. The statistics of the signal must be known or estimated for this filter to be designed, i.e., we must find R and $\underline{r}$.

The least mean–square adaptive algorithm [26,27] is an alternative to Weiner filtering that adapts the filter weights on a sample–by–sample basis. The filter gradually learns the proper correlations and adjusts its weights according to this knowledge. If the weights eventually converge to a steady state then the optimal Weiner filter weights are obtained.

The adaptive filter algorithm is designed in the same way as that for the Weiner filter, but the error criterion is modified and we use the notation $\underline{a}$ for the filter weights. The estimation error is

$$E[e_n^2] \quad = \quad E[(d_n - \underline{a}x_n)^2] \tag{86}$$

We minimize $E[e_n^2]$ by taking the partial derivative with respect to $\underline{a}$ and setting it equal to zero. This gives

$$R\underline{a} \quad = \quad \underline{r} \tag{87}$$

From which we have

$$\underline{a} \quad = \quad R^{-1}\underline{r} \tag{88}$$

The adaptive implementation comes about by solving

$$\frac{\partial E[e_n^2]}{\partial a_i} \quad = \quad 0 \ , \quad 1 \leq i \leq p \tag{89}$$

in an iterative manner using gradient descent techniques (see Figure 4). In the adaptive mode the filter parameter a_i is made time–dependent, $a(n)$, and is updated each time sample. Then

$$a(n+1) \quad = \quad a(n) + \Delta a(n)$$

where $\Delta a(n)$ is a correction term. It so happens that we can obtain the expression

$$a(n+1) \quad = \quad a(n) - \mu \ \frac{\partial E[e_n^2]}{\partial a} \tag{90}$$

which in turn is

$$a(n+1) \quad = \quad a(n) + 2\mu e_n x_n \quad . \tag{91}$$

The parameter μ is the gain constant.   This parameter is crucial to the filter since it regulates the speed and stability of the adaptive process. The filter weight changes occur at each iteration and are often calculated using gradient estimates.

There are many problems to consider in adaptive filter design.   We have only provided a brief introduction of the subject; for more detailed treatment of the subject, the reader is referred to references 26 and 27.

10.10.1   FEATURE EXTRACTION AND PATTERN RECOGNITION [9,10,15,28–31]

One of the goals of feature extraction is to reduce the size of the measured data set to as few parameters as possible, i.e., reduce the dimensionality of the data.   Achieving this goal facilitates the classification of the data and the detection of biomedical data in realtime as well as assisting in the recognition of perturbations in the data due to changes in physiological functioning or manipulations of a stimulus during an experiment.   Other applications are to simply recognize the presence or absence of an ERP waveform or a component within the ERP signal.   Investigators use the features to classify the data into clusters that have certain features in common. Nearly all procedures used in pattern recognition, the process of classifying a measured data set into mutually exclusive categories or classes, have come from hypothesis testing and statistical decision theory.

10.10.2   BAYESIAN DECISION THEORY

Bayesian decision theory is a fundamental statistical approach to the data classification problem.   This technique is based on the assumption that the decision problem is posed in probabilistic terms with a complete knowledge of all relevant probabilities.   The problem of classifying an arbitrary pattern $\underline{x}$ of unknown class is solved by using Bayes' rule of conditional probability.   An expected loss or conditional risk is defined for making a decision.   Another approach is to find the decision rule that minimizes the error rate of misclassification.   The decision rule is a mechanism for partitioning the feature space into regions, with each region consisting of samples from a particular class.   If there are only two classes then the decision rule divides the feature space into two regions.   If one knows the requisite conditional probability density functions, then the decision rule can be expressed using a discriminant function, often calculated using the Gaussian probability density functions. Special cases, such as when all features are statistically independent and have the same variance or when both classes have the same covariance matrix, lead to linear discriminant functions.   If the covariance matrices for the classes are different, then the discriminant function is quadratic in nature.

Unfortunately, we usually have no knowledge of the conditional probability densities for our feature vectors and the class it belongs to.   So Bayesian theory cannot be used.   One might use a training set to estimate the unknown probabilities and probability densities.   In practice this does not work well.   Often what is done is that one assumes the form of the probability density is known but that one or more parameters of the density are unknown and are to be estimated.   This is the parametric point estimation problem.

Nonparametric methods do not require knowledge of the form of the density, but these methods require a larger number of training samples.

The parametric estimation problem is frequently solved using maximum likelihood estimation or Bayesian estimation. These methods are discussed in many text books. We point out that sometimes these approaches can lead to complicated discriminant functions which may be difficult to implement.

10.10.3   LINEAR DISCRIMINANT FUNCTIONS

A different approach is to assume the form of the discriminant function and then use the training samples to estimate the values of its parameters. These parameters can be determined in various ways and have been regarded as the learning method. None of these procedures require knowledge of the underlying distributions; consequently, we can select any form of the discriminant we desire. One of the most popular discriminant functions is the linear function, because such functions can be implemented easily and are optimal in some cases if the underlying distributions are of the proper form.

Linear Discriminant Function of the Two–Class Case.

A linear discriminant function has the form

$$g(\underline{x}) = w_1 x_1 + w_2 x_2 + ... + w_p x_p + w_0 \tag{92}$$
$$= \underline{w}^t \underline{x} + w_0 \ .$$

The p–dimensional vector $\underline{w} = (w_1, w_2, ..., w_p)^t$ is called the weight vector and w_0 is the threshold weight. The decision rule corresponding to the discriminant function $g(\underline{x})$ is

$$\hat{\omega}(\underline{x}) = \omega_1 \quad \text{if } g(\underline{x}) > 0 \tag{93}$$
$$= \omega_2 \quad \text{otherwise.}$$

If $g(\underline{x}) = 0$, an arbitrary class can be assigned. The decision rule partitions the feature space into two regions, one where $g(\underline{x}) > 0$ and another where $g(\underline{x}) < 0$. Since $g(\underline{x})$ is a linear function in $\underline{x}$, the decision surface is a p–dimensional hyperplane and is given by

$$\underline{w}^t \underline{x} = -w_0 \tag{94}$$

The geometric properties of the hyperplane can be found in text books on pattern recognition. The weights of the discriminant function must be found before the decision boundary can be determined. To do this consider a change in notation. Let $\underline{y}$ denote the $(p+1)$ dimensional vector

$$\underline{y} = [x_1, x_2, ..., x_p, 1]^t \tag{95}$$
$$= [\underline{x}, 1]^t \quad .$$

Let $\underline{a}$ denote the $(p+1)$ dimensional weight vector

$$\underline{a} = [w_1, w_2, ..., w_p, w_0]^t \tag{96}$$
$$= [\underline{w}, w_0]^t \quad .$$

With this notation, the disciminant function can be written as

$$g(\underline{x}) = \underline{a}^t \underline{y} \quad . \tag{97}$$

The vector $\underline{y}$ is known as the augmented pattern vector.

One method for determining the weight vector $\underline{a}$ is known as the perceptron learning algorithm. This algorithm assumes that the training samples are linearly separable. A second algorithm is known as the least mean–squared–error procedure. The training samples for this second procedure cannot always be linearly separated. The training samples are said to be linearly separable if there exists a linear classifier that can classify them correctly. In other words, there exists a vector $\underline{a}$ such that

$$\underline{a}^t \underline{y}_i > 0 \text{ for every } \underline{y}_i \in \omega_1 \tag{98}$$

$$\underline{a}^t \underline{y}_i < 0 \text{ for every } \underline{y}_i \in \omega_2 \tag{99}$$

The perceptron algorithm leaves the weight $\underline{a}$ unchanged if it classifies the sample correctly but moves the weight vector in the "right" direction when it makes a mistake. Convergence of the algorithm is guaranteed if the training samples are linearly separable.

If the experimental design set is not linearly separable, then one might consider the least mean–square error (LSME) procedure. This algorithm uses all training samples, whereas the perceptron algorithm focuses only on the misclassified samples. Previously, we sought a weight vector $\underline{a}$ to give the inner products $\underline{a}^t \underline{y}_i > 0$. The least mean–square error approach tries to make $\underline{a}^t \underline{y}_i = b_i$, where the b_i are arbitrarily specified positive constants. We have now replaced the problem of solving a set of linear inequalities with a problem of solving a set of linear equations. The solution vector $\underline{a}$ will depend on the vector $\underline{b}$. For an arbitrary $\underline{b}$ there is no reason to believe that the LMSE solution yields a separating vector in the linear separable sense. But the solution should give reasonable results in both the (linearly) separable and nonseparable cases.

10.10.4   FISHER'S LINEAR DISCRIMINANT

This linear discriminant function is rather different from those above. Fischer's linear discriminant does not solve the classification problem. Its purpose is to reduce the dimensionality of the data from p–dimensions to one dimension and still retain class separability. More specifically, we want to find a linear transformation that maps the data from feature space into one dimension and this mapping should maximize some criterion function that serves as the indication of class separability.

As we reduce the dimensionality of the data set, we wish to maintain good class separability, i.e., the projected points for the same class should cluster together, but the two clusters should remain separated (presuming they are reasonably separated in the higher dimensional space). This means that the projected points should have a large difference between the sample means and have small within group variances.

The Fisher linear discriminant is defined as the linear function $z = \underline{w}^t \underline{x}$ (a scalar) for which the criterion function

$$J(w) \;=\; \frac{\underline{w}^t S_B \underline{w}}{\underline{w}^t S_W \underline{w}} \tag{100}$$

is maximized. The matrix S_B is the between–class scatter matrix, which is symmetric and positive semidefinite:

$$S_B \;=\; (\underline{m}_1 - \underline{m}_2)\,(\underline{m}_1 - \underline{m}_2)^t \tag{101}$$

where $(\underline{m}_1 - \underline{m}_2)$ is the difference vector between the sample mean vectors. The within–class scatter matrix is S_w and is given by

$$S_W \;=\; S_1 + S_2 \tag{102}$$

where

$$S_i \;=\; \sum_{\underline{x} \in X_i} (\underline{x} - \underline{m}_i)(\underline{x} - \underline{m}_i)^t \tag{103}$$

It can be shown that the $\underline{w}$ that maximizes J is

$$\underline{w} \;=\; S_W^{-1}\,(\underline{m}_1 - \underline{m}_2) \tag{104}$$

which is Fisher's linear discriminant, i.e., the linear function that maximizes the ratio of the between–class scatter to the within–class scatter. This reduces a p–dimensional problem to a one–dimensional problem. Fisher's linear discriminant can be used to reduce the dimensionality of the data via feature extraction and selection. The mapping cannot reduce the minimum achievable classification error rate.   In general, one must sacrifice some of theoretically attainable performance for the advantage of being able to work in a lower dimension.   However, if the conditional density functions are multivariate Gaussian with equal covariance matrices, then there need not be a sacrifice in performance.

Fisher's linear discriminant can be interpreted as a method for finding the hyperplane that partitions the feature space.   However, the discriminant only gives the direction (orientation) $\underline{w}$ of the hyperplane.   The discriminant cannot determine the position, i.e., the threshold weight.   This is why Fisher's discriminant does not solve the classification problem.

10.10.5   FEATURE SELECTION AND EXTRACTION

Feature selection is concerned with picking features from the measurement space, while selecting features from the transform space is known as feature extraction.   Feature selection involves the application of an algorithm to choose a subset of p—measurements (features) that maximize some criterion function.   Thus, an optimal selection can be achieved only by testing all possible sets of p features chosen from P measurements.   Even for a modest number of features the calculations required can be large since we must apply the criterion

$$\begin{bmatrix} P \\ p \end{bmatrix} \quad = \quad P!/(p!(P-p)!) \tag{105}$$

times.   In practice, suboptimal procedures are used [29–31].

For feature extraction, the problem is to find a p×p matrix W such that the derived features $\underline{y} = W\underline{x}$ maximize some criterion.   All feature extraction algorithms can be classified into two catgories.   The first category solves for the matrix W directly.   This method is similar to Fischer's linear discriminant method.   For the second category the data are transformed to another domain, for instance by the Fourier transform, the Hadamand transform, the Karhunen—Loeve expansion, etc.   Then features are selected from the transformed domain.

10.10.6   FEATURE SELECTION CRITERIA

The ultimate goal of a pattern recognition system is to properly classify the data with as low a classification error rate as possible.   Thus, an ideal criterion for feature selection is to minimize the error rate. Unfortunately, the error rate is difficult to evaluate.   To overcome this theoretical problem other criteria are chosen which vary monotonically with the error rate.   These criteria are selected for their ease of calculation and they should possess the following properties:

1.   Should be monotonically increasing or decreasing with the probability of error, or with the bound (upper or lower) on the probability of error.

2.   Should be invariant under one–to–one mapping.   This property is important since the probability of error is invariant under any transformation that possesses a one–to–one correspondence,

3.   All the selected features should be uncorrelated, and

4.   Satisfy the metric properties, i.e.,

i) $c(\omega_i,\omega_j : Y_m) > 0$ for $i \neq j$

ii) $c(\omega_i,\omega_i : Y_m) = 0$

iii) $c(\omega_i,\omega_j : Y_m) = c(\omega_j,\omega_i : Y_m)$

iv) $c(\omega_i,\omega_j : Y_m) \leq c(\omega_i,\omega_j : Y_{m+1})$

where Y_m denotes a subset of m features, ω_i denotes class ω_i and $c(\cdot)$ denotes a criterion.

Note that all the above properties are not necessary but are desirable. So far for the general case, none of the existing criteria satisfy all of the above criteria.

Two popular criteria are based on Fisher's ratio and the Karhunen–Loeve expansion [10,15].

10.10.7   PERFORMANCE ESTIMATION

After the classifier is designed, one needs to evaluate its performance to compare the design with competing approaches.   Consider the error rate as the performance measure.

Four popular empirical approaches that count the number of errors when testing the classifier with a test data set are:

(1)   The Resubstitution Estimate.   In this procedure, the same data set is used for both designing and testing the classifier.   Experimentally and theoretically this procedure gives a very optimistic estimate, especially when the data set is small.   Note, however, that when a large data set is available, this method is probably as good as any procedure.

(2)   The Holdout Estimate.   The data is partitioned into two mutually exclusive subsets in this procedure.   One set is used for designing the classifier and the other for testing.   This procedure makes poor use of the data since a classifier designed on the entire data set will, on the average, perform better than a classifier designed on only a portion of the data set.   This procedure is known to give a very pessimistic error estimate.

(3)   The Leave–One–Out Estimate.   This procedure assumes that there are n data samples available.   Remove one sample from the data set. Design the classifier with the remaining (n–1) data samples and then test it with the removed data sample. Return the sample removed earlier to the data set. Then repeat the above steps, removing a different sample each time, for n times, until every sample has been used for testing. The total number of errors is the leave–one–out error estimate. Clearly this method uses the data very effectively. This method is sometimes referred to as the Jack Knife method.

(4)   The Rotation Estimate.   In this procedure, the data set is partitioned into n/d disjoint subsets, where d is a divisor of n.   Then, remove one subset from the design set, design the classifier with the remaining data and test it on the removed subset, not used in the design.   Repeat the operation for n/d times until every subset is used for testing.   The rotation estimate is the average frequency of misclassification over the n/d test sessions.   When d=1 the rotation method reduces to the leave–one–out methods.   When d=n/2 it reduces to the holdout method where the roles of the design and test sets are interchanged.   The interchanging of design and test sets is known in statistics as cross–validation in both directions.   As we may expect, the properties of the rotation estimate will fall somewhere between the leave–one–out method and holdout method.   The rotation estimate will be less biased than the holdout method and the variance is less than the leave–one–out method.

Another approach to performance estimation is to assume a form for the distribution function with unknown parameters. This is called the parametric approach. The previous methods do not require a knowledge of such distributions. In the parametric method the parameters have to be estimated from the available training samples. This technique leads to the estimation of the mean values and covariance matrices. A common criterion is the Mahalanobis distance measure.

There are other factors that affect the performance of the classifier, such as the size of the training set, whether or not the training and testing set sizes are equal and others [9,29–31].

Applications to EEG and ERP data may be found elsewhere [9,10,15, 28–31]. A major use for these techniques is the classification of single–trial ERPs, i.e., individual event–related potentials elicited by the occurrence of a single stimulus. In these cases no signal averaging is used. Consequently, the signal–to–noise ratio is low. Feature selection is crucial if one is to minimize the classification probability of error [15].

10.9.8   SYNTACTIC METHODS [32]

Syntax is the discipline of sentence structure and the rules that govern the arrangement of words in sentences for a language. Syntactic pattern recognition uses this concept. Such systems use feature selection and extraction, the assignment or selection of basic elements that are analogous to words, and the selection of rules that are used to connect the basic elements. This process is called syntax analysis. If the signal to be analyzed is a time waveform, such as a saw tooth (triangular) waveform, then we might have two basic elements, the rising slope and the falling slope of the saw–tooth waveform.   Let us add a third element, such as a horizontal segment (dc level) placed after the falling slope segment, then we can have some rules. For example, a falling slope must follow a rising slope, a rising slope must follow a horizontal segment and a horizontal segment must follow a falling slope.   For this simple example we allow no exceptions.   The features of the waveform are these basic elements.   The waveform may be recognized by using these features and the rules that govern the assembly of these features. Artifacts in the signal waveform can be defined and recognized in the same manner as can signal perturbations due to noise.

If the application is more complicated, then we have problems.   We must understand the signal generation mechanism to identify features and construct rules.   Both of these processes have defied automation.   In the recognition phase we need to parse the waveform and such algorithms are not yet very sophisticated when dealing with difficult problems.   But, nonetheless, these systems are finding use in biomedical application.

10.11   EXPERT SYSTEMS [4,33,34]

One application area of artificial intelligence (AI) is knowledge–based expert systems, some of which were first developed to assist medical decision making. The conventional algorithm–based computer programs often do not apply to these areas since algorithms may not exist for these problem domains.   Instead, heuristic knowledge is used by the human expert to solve the problem.   This heuristic approach resists precise description and rigorous analysis.

Contemporary methods of symbolic and mathematical reasoning, which have limited applicability to the area of expert systems, do not provide the means for 1) representing knowledge, 2) describing problems at multi–levels of abstraction, 3) allocating problem–solving resources, 4) controlling cooperative processes, and 5) integrating diverse sources of knowledge for problem solution inference.   The expert system approach depends primarily on the capacity to manipulate problem descriptions and to apply relevant pieces of knowledge selectively.   These systems, depending on the application problem domain, are categorized into several types such as interpretation, prediction, diagnosis, design, planning, monitoring, debugging, repair, instruction, and control systems.

Expert knowledge consists of two parts: 1) symbolic descriptions that characterize the definitional, taxonomical, and empirical relationships in a domain, and 2) the procedures for manipulating these descriptions.   To achieve a high–level of performance, a human expert's skills also need to be understood well and included in the knowledge base.

Expert system architectures vary widely depending on their applications. The current techniques and principles of expert systems are based mostly on the relatively small number of early expert systems.   Rule–based knowledge representation and system structure constitute the major framework for the current expert system technologies.   One application is known as semantic techniques that may be used to analyse waveforms.

Another application of expert system technology is in the building of medical consulting systems designed as aids to medical decision making.

Most errors made by the clinician are caused by omissions involved in the diagnostic process. The computer can provide reliable diagnostics by an exhaustive consideration of all the possibilities and all the patient's relevant data. Clinical medicine has also been a fertile area for the study of cognitive processes, since diagnosis as a cognitive process has been studied extensively.

The medical decision making process involves three parts: data gathering, diagnosis, and treatment. Data gathering consists of obtaining the patient's history and clincial and laboratory data. The clinical data include the symptoms, which are subjective sensations reported by the patient, and signs, which are observable by the physicians. The laboratory data may include the results of tests. Diagnosis is the process of using the data to determine the illness. Gathering information, diagnosing the disease, and deciding on a treatment regimen constitute a consultation. Often, medical knowledge is incomplete, and it may not be possible to determine the cause of the patient's disease. In these cases treatments must be based on the empirical associations of disease characteristics. Provisions for the use of indirect data or additional information gathering must be included in the design of the computer system.

Historically, statistical analysis and pattern recognition using a discriminant function based on the Bayesian decision theory have been used for computer analysis of some diagnostic problems. The appeal of the statistical method is that the decisions are optimal within the basis of a given criterion. However, the statistical approach is unsuitable for many medical problems, because of the various assumptions and simplifications that must be made. These assumptions cannot be suitably validated and the a priori and conditional probabilities required in the analysis are usually not available. Medical diagnostic problems can be viewed as a problem of hypothesis formation. The diagnostic task involves using clinical findings to form a disease hypothesis.

10.11.1 MYCIN [33]

The Mycin system was developed to provide consultative advice on diagnosis and therapy for infectious diseases. Its knowledge–base is encoded in production rules. The internal form of the rules is encoded in LISP. Each rule represents domain specific knowledge. The knowledge is highly stylized with fixed syntax of premise–action form and limited primitives. The system also provides translation capabilities to explain the lines of reasoning. Originally, it used about 200 rules. The inexactness and incompleteness of the knowledge are handled by the certainty factor scheme. The certainty factors are considered as part of the knowledge and are used in the inference of a conclusion. A certainty factor is a measurement of the association between the premise and the action clauses of each rule. The knowledge consists of two parts: 1) the attributes, objects, and values, which form a vocabulary of domain–specific conceptual primitives, and 2) the inference rules expressed in terms of these primitives. The inference engine refers to the mechanism used to draw conclusions based on the rules in the knowledge base and the data for the current case. Rules are invoked in a backward chaining fashion that results in an exhaustive depth–first search of an AND–OR goal tree. Mycin considers all the possibilities. The total weight of the evidence about a hypothesis is tested with an empirically selected threshold. If the weight falls below the threshold then it asks for information from the user. The expert may inspect the reasoning chains and add and modify any rules or clinical parameters required to augment and repair the medical knowledge of Mycin. Formal evaluations of the Mycin system have shown that Mycin compares favorably with experts in infectious diseases for diagnosing and selecting therapy for patients with bacteria and meningitis. The system is not used in medical wards because of its incomplete knowledge of the full spectrum of infectious diseases.

10.11.2 CASNET [34]

The causal association network (CASNET) is a computer system developed at Rutgers University for performing medical diagnosis. The major application of CASNET has been in the treatment of glaucoma. The disease is represented as a dynamic process that can be modeled by a network of causally linked pathophysiological states. The system diagnoses a patient by determining the pattern of pathophysiological causal pathways present in the patient and identifying this pattern with a disease category. The CASNET model consists of three planes of knowledge: a plane of the pathophysiological states, a plane of observations, and a plane of classification tables. The plane of pathophysiological states is the major part of the model. The nodes in this plane represent elementary hypotheses about the disease process, and the arcs represent causal connections between two elementary hypotheses. A confidence factor or forward weight, which is a number scaled between 1 and 5, is associated with each link. The plane of observations contains nodes representing evidence gathered from the patients such as signs, symptoms, and laboratory tests. A classification table defines a disease as a set of confirmed or denied pathophysiological states. It also contains a set of treatment statements for that disease. Reasoning in CASNET is designed to maximize the likelihood of finding the pathways which are defined as a disease in conjunction with causally related pathological states, given a set of signs, symptoms, and test results. A confidence factor for each node in the causal net, i.e., a status, is calculated by using the user input patient data as well as associated confidence factors with the tests and weights associated with the causal arcs. The status value is affected both by the results of its associated tests and by the states around it. A general algorithm is used to propagate these weights on a state both in the forward and backward direction. A state is marked, if the value is below the threshold. The system employs a strategy to select the next question based on the cost of the test and the likelihood that it will lead to the confirmation or denial of a state. After all the symptoms and findings have been entered, the classification tables are used to determine diagnoses and treatments. The CASNET adopts a strictly bottom–up approach to the problem of diagnosis, working from the tests, through the causal pathways, to a diagnosis. This model also provides a convenient way of following the process of a patient's disease by using the causal net to view the disease progression both in a backward and forward direction along the pathways. The program's performance has been evaluated by ophthalmologists and is considered close to the expert level.

10.11.3 INTERNIST [34]

INTERNIST is a consultation program in the domain of internal medicine developed at the University of Pittsburg. The system is presented with a list of manifestations of disease in a patient and it attempts to form a diagnosis. Using the information presented during the consultation, the program tries to discriminate between competing disease hypotheses. Thus, this system attempts to solve a hypothesis. INTERNIST's knowledge is represented in the form of a disease tree, or disease taxonomy. The top–level classification in this tree is by organs. A disease node's offsprings are refinements of that disease, terminal nodes being individual diseases. The disease hierarchies are predetermined and fixed in the system. Each disease in the tree is associated with relevant manifestations, and several relationships are superimposed on the disease tree to capture causal, temporal, and other association patterns among diseases. The disease tree and its associated manifestations are constructed and maintained separately from the diagnostic program. A list of manifestations is computed for each nonterminal node by taking the intersection of the manifestation lists of that node's offspring. In this way, the manifestations associated with a nonterminal disease node are also associated with every node beneath it in the tree. As well as providing storage economy for this hierarchy, this information is used during consultation for selecting disease areas on which to focus.

At the beginning of a consultation, a list of manifestations is entered, which evokes one or more nodes of the disease tree. A model is created for each evoked disease node. The model consists of lists which are: observed manifestations that this disease cannot explain, observed manifestations that are consistent with the disease and manifestations that should be present if the disease is the correct diagnosis but have not been observed in the patient.

A diagnosis corresponds to a set of evoked terminal nodes that account for all of the symptoms. In general, in the beginning, very few terminal nodes will be evoked, so the program must ask for further information. To get this additional information, the program will focus on a disease area and formulate a problem. Disease models are scored and various disease models are considered as problems. Competing candidates are considered in turn until a diagnosis is made. INTERNIST has a knowledge base of about 500 diseases of internal medicine (75% complete). It has displayed expert performance in complex cases involving multiple diseases.

Expert systems is a natural expansion area for biomedical signal processing professionals since they are already familiar with the accepted hypotheses within their application area. One approach is to design an expert system to confirm or deny the prevailing hypotheses by using data gathered from experiments or patients. The expert system design would conceivably formalize the rules and gather the knowledge necessary to confirm the critical hypotheses. One objective of such systems would be to teach new professionals in a more economically concise manner because the rules and knowledge base would be more precise and better organized than that which presently exists. But the object is not to replace the clinician, rather to provide assistance.

10.12 MOLECULAR COMPUTING

Can we devise a computer that uses molecules to sense, transform and provide output signals? A coordinated, interdisciplinary, and international effort was initiated by NSF about four years ago to examine the problems of building such an information processing system [35–36]. The technical areas that are projected to be needed in this endeavor are biosensors, protein engineering, recombinant DNA technology, polymer chemistry, and artificial membranes. The disciplines needed will include biologists, chemists, physicists, computer scientists, and electrical engineers. The objective is to design and build computers made of proteins and other large molecules rather than silicon integrated circuits.

The analogy Conrad [36] has used to illustrate molecular computing is the immune system. Imagine you cut your finger with an old fishing knife. Harmful bacterial enter your bloodstream. As in the movie "Fantastic Voyage" white blood cells recognize an invader as it rushes by. A defensive reaction follows. Chemical reactions occur between and within cells. No electrical signals from the brain are required. The finger swells providing evidence of these chemical and cellular events. The immune system has worked like a parallel, adaptive computer. A chemical input on a cell membrane causes the cell to act. Further, the immune system acts with intelligence. If the bacterium is determined to be a foreign body, then more antibodies may be produced and released into the blood stream. Other decisions may be rendered as well.

This anecdote illustrates the possibility of molecular computing. Can this concept be extended and applied to build pattern recognition molecular computers?

The research at the present time suggests that the new moleculear computers, if and when they become a product, are not meant to compete with conventional computers as we know them today. The new approach is imagined to be context dependent. Inputs will be processed as dynamic physical structures, not bit by bit. Protein molecules recognize other molecules in their environment by reacting to their shape. This recognition process will be used to effect computation rather than using the present day digital switches that pass or inhibit the flow of electrons depending on whether a threshold level is exceeded or not.

Molecular computers are envisioned to benefit from context dependence since they will be better suited for processing patterns of sensory inputs such as light, temperature, pressure. Artificial vision and touch will rely on the action (computation) or protein molecules in their physiochemical environment. The computer "program" will be implicit in the protein's structure. Physical recognition will be the basis of computation. The first such computers are envisioned as sensors that see patterns, feel surfaces, and sense chemical gradients. These sensors will probably be interfaced initially to conventional computers.

Presently two approaches are being considered for fabricating these new computer systems. One seeks to duplicate certain biological system capabilities, such as object recognition, learning, and use of parallel computation. The other is considering the creation of molecular electronic components that might lead to smaller, faster, less expensive computers. For either approach to be successful, molecules will have to be linked into long chains to perform the necessary computation. Naturally occurring biological systems are being studied as models, while synthetic structures are also being experimented with.

Present computers are structurally programmable, the computation is done symbolically, and the program is developed by a human who understands the algorithm to be performed. In the molecular computer we will find the simple switching devices replaced by protein enzymes. Patterns will be processed physically and dynamically rather than symbolically and passively. The program will be represented implicitly rather than explicitly in the structure of the proteins and the system in which they are contained. Structural programming as we know it will not exist. The programming will depend on evolution by variation and selection.

Each protein enzyme recognizes a specific type of molecule, called the substrate and switches its state by making or breaking a precisely selected chemical bond. This switching action is a complicated computation, which is done in a geometric rather than a logical mode. The environment and other molecules affect the recognition process by affecting the shape and motion of the molecule. The recognition process is tactile. Consequently, the enzyme can be switched to different shape states, allowing for control and memory storage at the molecular level.

The protein's shape depends upon the interactions among amino acids and how the chain is folded. The fold determines the computational function. This cannot be programmed, which is good since small changes in the amino acid sequence may produce a small change in folded shape and a small change in function. But the relationship between the amino acid sequences and the protein shape is not rigid as it is between a computer code and its function. Therefore, molecular computers will be able to learn through evolution.

According to Conrad [36] molecular computer design must solve the problem of how to use the geometry recognition capabilities of enzymes to process nontactile input signals, such as photons or electronic pulses. Researchers presently think that the enzymes will have to be placed in a processor that will be able to convert the input signal into a molecular and physiochemical form that enzymes can recognize. These processors will have a receptor layer that will transduce input signals into a pattern in the second layer. If the second layer is substrate molecules in solution, then this pattern will be in the form of a reaction–diffusion pattern in the substrate molecules. Enzymes in the third layer will read local concentrations of the reactions and control the output signal of the processor. The program of the processor will be determined by the manner in which the input signals are converted and how the enzyme readout process will work. To make this work will require considerable skill in manipulating the evolutionary process of the input molecular conversion and output enzyme processes.

Various design approaches are presently being considered and some are being experimented with, such as gene and protein engineering, immunological techniques, polymer chemistry, membrane engineering, and others. Biosensors might be considered a beginning approach since these sensors transform input information, perhaps in chemical form, into electrical output. There are sensors for sensing glucose, oxygen, etc. Such sensors might consist of detector enzymes on a membrane and a transducer to provide an electrical or optical output in response to the enzyme reaction at the input.

It may be a century before molecular computers will reach the market place, but this is an exciting new area for biomedical signal processing. The field of molecular computing is presently limited by materials and analytical methods for design. Some feel that a closer coupling between computing and biology will be needed to achieve a molecular computer. Biosensors seem the logical first step to take [35–36].

10.13 NEURAL NETWORKS [37,38]

The concept of neural networks dates to the 1940s. The goals of this line of research are perhaps two–fold: first, to understand human cognitive processing and second, to model new computer architectures after human brain functioning.

Humans are smart. But why? We certainly do many thinngs more slowly and with less efficiency than some machines. But other tasks are quite easy for us and we are faster than any existing machine. For example, we recognize people easily, not only their appearance but also their voice. We seem quite adept at considering many pieces of information and/or constraints simultaneously. Further, we can recall and make use of seemingly unrelated material to solve a problem. We appear to perform these tasks using neural networks that compute data both serially and in parallel.

Natural language processing is performed quite easily by humans. We consider syntax to help us determine the meaning (semantics) of a message. But we can also use the semantics to interpret the syntax. We use constraints to recognize words in speech or in the printed text which have been corrupted by "noise".

Motor control of our limbs or of our speech mechanism appears to use multiple constraints and anticipation. As we speak we anticipate the forthcoming articulatory movements required of our jaw, lips, tongue. Such motions are also constrained. We also seem to perform the necessary control of many muscles in parallel. Some elementary models of the neural speech production and auditory listening processes are available. But these models are still primitive. Still the idea is to build neural networks based on such models. These models must account for memory storage and recall, pattern recognition, associative memory mechanisms, and other facors. The architecture must consider how the elements of the model are interconnected to facilitate learning. Research along these lines is expected to influence network architectures, learning algorithms, speech recognition and synthesis, parallel computer design, and associative memory concepts.

Existing computer systems execute the programmed instructions sequentially, while neural networks examine many hypotheses simultaneously. To do this, these networks are made up of parallel nets of many computational elements connected by links that have variable weights. The computational elements are typically analog and non–linear, thus contrasting with our present digital computers, but reminiscent of the analog computers of twenty years ago. A simple node might sum several inputs and threshold (hardlimit) the sum. Other possible nodes might integrate, differentiate, or perform complex mathematical operations.

Existing neural net models have a network topology, node characteristics and a set of training rules. The rules specify the start–up procedure, such as initial weights and then govern how the weights are to be changed during the learning process.

The adaption or learning ability of these neural nets is a feature that could be exploited to great advantage in biomedical signal processing. This paper has repeatedly stressed the variability of biomedical data depending on the subject. Another area with similar problems is speech recognition independent of the speaker. Both of these fields usually have limited training data but want to use the algorithm or devices with new individuals or in new environments. The ability of the neural network to adapt would provide a formal mechanism that could compensate for variabilities in the subject's EEG or ERP responses or his speech. Our present pattern recognition approaches to these variability problems typically use all of the initial training data in the design stage before new data is processed. Often the new data is not used to update the training of the pattern recognition algorithm. Neural networking appears to provide an alternative approach. Further, the classifiers for these networks can be non–parametric and make use of weaker assumptions about the statistical distributions of the data.

We have already described how the traditional pattern recognition approaches are presently applied to biomedical data. An adaptive neural net classifier may have parallel data connections to the first stage (as opposed to a single stage input), thus accommodating a feature vector. The input data (feature vector) may be binary or analog. The succeeding network stages might then do a k–means clustering or vector quantization for selecting or enhancing the maximum input. Some algorithms already exist that perform these tasks with learning. To date, the performance of these neural networks has not surpassed the ability of existing conventional pattern recognition algorithms. But this status may soon change.

10.14   CONCLUSION

This paper has attempted to provide the reader with a view of the more traditional approaches to biomedical signal processing, such as pattern recognition, signal averaging, filtering, spectral analysis and others. The traditional goals in this field have been the monitoring and interpreting of the signals. This often ecompasses modeling of the biological system as well as modeling of the signals measured from such a system. Newer areas of research include human–computer interaction for the possible development of neural control devices. Molecular computing using protein molecules instead of silicon have only just begun to be considered as a new way to achieve certain pattern recognition tasks. Neural networking is about 40 years old, but has made rapid advances in recent years with the invention of specific learning algorithms that can be tested or simulated using existing computers. Throughout the paper we have attempted to mention areas and problems that need further research. We hope the reader finds these suggestions of interest.

10.15 REFERENCES

1.   J.B. Bronzino, V.H. Smith, and M.L. Wade, "Evolution of the American Health Care System and Economic and Ethical Implications," Chapter I in Technology and Medicine, J.D. Bronzino, V.H. Smith, and M.L. Wade (Eds.) (Trinity College, Hartford, CT, 1985), pp. 7–19.

2.   A.M. Grass and F.A. Gibbs, "A Fourier Transform of the Electroencephalogram," Jour. Neurophysiol., 1 (1938), 521–526.

3.   E.O. Attinger, "Impacts of the Technical Revolution on Health Care," IEEE Trans. on Biomed. Engr., BME–31 (Dec. 1984), 736–743.

4.   M.H. Eckman, "The Role of Computers in Clinical Decision Making," Chapter V in Technology and Medicine, J.D. Bronzino, V.H. Smith, and M.L. Wade (Eds.) (Trinity College, Hartford, CT, 1985), pp. 58–68.

5.   G.E. Loeb, "Neural Prostheses: Interfacing with the Brain," Chapter III in Technology and Medicine, J.D. Bronzino, V.H. Smith, and M.L. Wade (Eds.) (Trinity College, Hartford, CT, 1985), pp. 42–49.

6.   F.F. Offner, "Bioelectric Potentials – Their Source, Recording, and Significance," IEEE Trans. on Biomd. Engr., BME–31 (Dec. 1984), 863–868.

7.   F.M. Galioto, "Cardiovascular Assist and Monitoring Devices," Chapter II in Technology and Medicine, J.D. Bronzino, V.H. Smith, and M.L. Wade (Eds.) (Trinity College, hartford, CT, 1985), pp. 20–41.

8.   D.G. Childers, "Evoked Responses: Electrogenesis, Models, Methodology and Wavefront Reconstruction and Tracking Analysis," Proc. IEEE, 65 (May 1977), 611–626.

9.   D.G. Childers, P.A. Bloom, A.A. Arroyo, S.E. Roucos, I.S. Fischler, T. Achariyapaopan, and N.W. Perry, Jr., "Classification of Cortical Responses Using Features from Single EEG Records," IEEE Trans. on Biomed. Engr., BME–29 (June. 1982), 423–438.

10. D.G. Childers, I.S. Fischler, N.W. Perry, Jr., and A.A. Arroyo, "Multichannel, Single Trial Event Related Potential Classification," IEEE Trans. on Biomed. Engr., BME–33 (Dec. 1986), pp. 1069–1075.

11. D.G. Childers, "Talking Computers: Replacing Mel Blanc?", Comp. Mech. Engr., 6 (Sept. 1987), pp. 22–31.

12. D.G. Childers, and K. Wu, "Some Factors Responsible for Quality, Intelligibility and Naturalness of Synthetic Speech," Jour. Acoust. Soc. Am., to appear.

13. N.W. Perry, Jr. and D.G. Childers, The Human Visual Evoked Response, Method and Theory, (Charles C. Thomas, Springfield, IL, 1969).

14.   J. Aunon, C.D. McGillem, and D.G. Childers, "Signal Processing in Evoked Potential Research, I. Averaging and Modeling, "CRC Critical Reviews in Bioengineering, 5 (July. 1981), 323–367.

15.   C.D. McGillem, J. Aunon, and D.G. Childers, "Signal Processing in Evoked Potential Research, II. Filtering Techniques and Pattern Recognition," CRC Critical Reviews in Bioengineering, 6 (Oct. 1981), 225–265.

16.   D.G. Childers, J. Aunon, and C.D. McGillem, "Signal Processing in Evoked Potential Research, III. Spectrum Analysis: Prediction and Extrapolation," CRC Critical Reviews in Bioengineering, 6 (Sept. 1981), 133–175.

17.   C.D. Dawson, "Cerebral Responses to Electrical Stimulation of Peripheral Nerve in Man," Jour. Neurol., Neurosurg. Psychiat., 10 (1947), 134–140.

18.   D.G. Childers and A. Durling, Digital Filtering and Signal Processing, (West Publishing Co., St. Paul, 1975).

19.   B.H. Jansen, "Analysis of Biomedical Signals by Means of Linear Filtering," CRC Critical Reviews in Bioengineering, 12 (1981), 343–392.

20.   P.Y. Ktonas, "Automated Analysis of Abnormal Electrencephalograms," CRC Critical Reviews in Bioengineering, 9 (1982), 39–97.

21.   D.G. Childers, Modern Spectrum Analysis, (IEEE Press, New York, 1978).

22.   S.B. Kesler, Modern Spectrum Analysis II, (IEEE Press, New York, 1986).

23.   S.L. Marple, Jr., Digital Spectral Analysis, (Prentice–Hall, Inc., Englewood Cliffs, NJ, 1987).

24.   S.M. Kay, Modern Spectral Estimation, (Prentice–Hall, Inc., Englewood Cliffs, NJ, 1987).

25.   A. Isaksson, A. Wennberg, and L.H. Zeterberg, "Computer Analysis of EEG Signals with Parametric Models, Proc. IEEE, 69 (Apr. 1981), 451–461.

26.   B. Widrow and S.D. Stearns, Adaptive Signal Processing, (Prentice–Hall, Inc., Englewood Cliffs, NJ, 1986).

27.   S. Haykin, Adaptive Filter Theory, (Prentice–Hall, Inc., Englewood Cliffs, NJ, 1986).

28.   D.G. Childers, N. Perry, Jr., I.A. Fischler, T.L. Boaz, A.A. Arroyo, "Event–Related Potentials: A Critical Review of Methods for Single–Trial Detection," CRC Critical Reviews in Biomedical Engr., 14 (1987), 185–200.

29.   J.M. Moser and J.I. Aunon, "Classification and Detection of Single Evoked Brain Potentials Using Time–Frequency Amplitude Features," IEEE Trans. Biomed. Engr., BMME–33 (Dec. 1986), 1096–1106.

30.   D.G. Childers, "Signal–Trial Event–Related Potentials: Statistical Classification and Topography," Chapter 14 in Topography Mapping of Brain Electrical Activity, F.H. Duffy (Ed.) (Butterworths, Boston, 1986), pp. 255–293.

31.   T. Achariyapaopan and D.G. Childers, "Optimum and Near Optimum Feature Selection for Multivariate Data," Signal Proc., 8 (1985), 121–129.

32.   R.G. Shiavi and J.R. Bourne, "Methods of Biological Signal Processing," Chapter 22 in Handbook of Pattern Recognition and Image Processing, T.Y. Young and K–S Fu (Eds.) (Academic Press, Inc., Orlando, FL, 1986), pp. 545–568.

33.   B.G. Buchanan and E.H. Shortliffe, "Rule–Based Expert Systems: The MYCIN Experiments of The Stanford Heuristic Programming Project" (Addision–Wesley Publ. Co., Reading, MA, 1984).

34.   A. Barr and E.A. Feigenbaum (Eds.), The Handbook of Artificial Intelligence, Vol. 2 (William Kaufman, Inc., Los Altos, CA, 1982), pp. 175–222.

35.   M. Conrad. "On Design Principles for a Molecular Computer," Comm. ACM, 28 (May 1985), 464–480.

36.   M. Conrad. "The Lure of Molecular Computing," IEEE Spectrum, 23 (Oct. 1986), 55–60.

37.   D.E. Rumelhart and J.L. McClellan (Eds.), Parallel Distributed Processing, Vol. 1: Foundation (A Bradford Book, MIT Press, Cambridge, MA, 11986).

38.   R.P. Lippman, "An Introduction to Computing with Neural Nets," IEEE Acoust., Speech, Signal Proc. Magazine, 4 (Apr. 1987) 4–22.

39.   D.S. Barth, W. Sutherling, J. Engel, and J. Beatty, "Neuromagnetic Evidence of Spatially Distributed Sources Underlying Epileptic Form Spikes in the Human Brain," Science, vol. 223, 1984, pp. 293–296.

40.   S.J. Williamson and L. Kaufman, "Biomagnetism," J. Magnet. and Magnetic Materials, vol. 22, 1981, pp. 129–202.

41.   J. Beatty, D.S. Barth, and W. Sutherling, "Magnetically Localizing the Sources of Epileptic Discharges Within the Human Brain," Naval Res. Rev., Vol. XXXVI, 1984, 20–28.

11
POLARIZATION AS A RADAR DISCRIMINANT

A. MACIKUNAS AND S. HAYKIN
Communications Research Laboratory
McMaster University
Hamilton, Ontario

11.1 INTRODUCTION

The polarization of a radar wave is a valuable discriminant that can be used for enhancing the radar detection, tracking, identification and classification performance. This chapter primarily deals with the improvements possible in detection, discrimination and tracking (radar navigation). For more information on using polarization information for identification and classification the interested reader is referred to Copeland (1960), Daley (1982), Ezquerra (1984), Corsini, et al. (1984) and Russell, et al. (1987). Polarization has found application in conventional marine navigation and surveillance radars, as well as more specialized applications such as ice detection and remote sensing radars. This chapter contains an up–to–date survey of polarization utilization in the above areas.

We begin by describing the polarization nature of a radar wave and the mathematical representation of polarized waves. Also included are the measurement requirements and the associated errors for obtaining polarization information about targets. Polarization descriptors such as the polarization scattering matrix and the Stokes vector, as well as their relationships to the Poincaré sphere and polarization chart projections will be described. Within this above framework, target and clutter discriminants based on optimal polarization, Stokes vector/matrix, polarization vector processing, and the Hermitian coherency matrix will be compared for coherent or partially–coherent radars. Other techniques suitable for non–coherent radars such as video addition, polarization amplitude ratios, and pseudo–coherent detection will also be described. Lastly, some thoughts will be presented on the reduction or cancellation of residual system– or measurement–related polarization errors.

Radars operate on the principle that objects tend to scatter incident electromagnetic radiation back in the direction of the radar receiving antenna. In the past, the performance of radars has been improved by any of the following schemes: judicious choice of operating frequency, increased power, narrow antenna beamwidth, compression of the transmitted pulse, doppler shift of the echo, etc. In virtually all conventional radars the state of polarization of the received echo is ignored. The most common situation is that a given radar transmits and receives only one fixed polarization; for example, linear horizontal. It is appreciated that this single polarization radar should have its transmitting polarization and receiving polarization matched. Sinclair (1950) formulated this matching condition for elliptically–polarized waves using a polarization vector approach. But, it is impossible to deduce the state of polarization of a radar wave using only one receive polarization.

The measurement of the polarization of the received echo requires a radar equipped with two coherent receivers connected to an antenna with two orthogonally polarized feeds (for example, linear horizontal and linear vertical, or right–hand circular polarization (RHCP) and left hand circular polarization (LHCP)). The associated extra hardware requirement in terms of both cost and complexity and the general lack of knowledge of the potential benefits of polarization have slowed advance in this area. More lately, however, decreasing hardware costs, and the failure of conventional radar discriminants to solve certain difficult problems have made polarization–sensitive radar (or polarimetric radar) a topic of much interest. Recent surveys of polarization utilization in radar have been compiled by Cloude (1983), Giuli (1986) and by Macikunas and Haykin (1986).

The mathematical framework for the description of polarized waves and associated phenomena has been well known for decades in the field of optics (Born and Wolf, 1980), and in electrical engineering (Ko, 1962). We will begin our study by describing the polarization scattering matrix and the Stokes vector in the next section.

11.2   MEASUREMENT AND DESCRIPTION OF POLARIZATION

11.2.1   POLARIZATION SCATTERING MATRIX

When electromagnetic waves from a radar strike a target or clutter–producing environment, an interaction takes place scattering the electromagnetic wave. The component of the radiation that is scattered back in the direction of the radar is usually polarized differently from the illuminating wave. Additionally, different target and clutter types possess different polarization characteristics, often varying with target aspect and illuminating frequency. Conventional radars can, at best, select a single pair of transmitting and receiving polarizations, i.e., transmit either linear horizontal (H) or linear vertical (V), and receive both H and V. The received signal voltage consists of the magnitude and phase angle of the radar echo, where the magnitude is proportional to the radar cross section (RCS). The complex RCS can be expressed in terms of the polarization scattering matrix (Wiesbeck, and Riegge, 1987). Specifically, we write

$$[\underline{\sigma}] = \text{the complex radar cross section}$$

$$= \begin{bmatrix} \underline{\sigma}_{HH} & \underline{\sigma}_{HV} \\ \underline{\sigma}_{VH} & \underline{\sigma}_{VV} \end{bmatrix} \tag{1}$$

where $\underline{\sigma}_{k\ell}$ are complex RCS with amplitude and phase. Moreover, the polarization scattering matrix may be written as follows:

$$[\underline{S}] = \begin{bmatrix} |S_{HH}| \exp(j\theta_{HH}) & |S_{HV}| \exp(j\theta_{HV}) \\ |S_{VH}| \exp(j\theta_{VH}) & |S_{VV}| \exp(j\theta_{VV}) \end{bmatrix} = \frac{1}{4\pi R^2} \begin{bmatrix} \underline{\sigma}_{HH} & \underline{\sigma}_{HV} \\ \underline{\sigma}_{VH} & \underline{\sigma}_{VV} \end{bmatrix} \tag{2}$$

Note that the polarization scattering matrix and the complex radar cross section can be related by a simple scaling factor, depending on the radar to target distance, R.

The polarization scattering matrix describes the radar scattering process in magnitude, phase and polarization. The scattering matrix can be used to predict the magnitude and phase of a single received polarized wave, given an arbitrary transmitted polarization; i.e., the scattering matrix is actually independent of the polarizations used for its measurement. Measurement of the scattering matrix requires a coherent radar that can provide the amplitude and phase of echo returns at two orthogonal polarizations simultaneously for two different transmit polarizations. These two transmitted pulses should have orthogonal polarization and can be transmitted in succession, for example, on a pulse–to–pulse basis, see Huynen (1965). Huynen also shows that the scattering matrix can be determined using six successive polarizations without the need for dual polarization or coherency. Unfortunately, this technique is best suited to laboratory conditions. In Equation (3) below, a polarization scattering matrix is shown for the case of linear vertical and horizontal transmission/reception:

$$\begin{bmatrix} E_H \\ E_V \end{bmatrix}_{scattered} = \begin{bmatrix} |S_{HH}|\exp(j\theta_{HH}) & |S_{HV}|\exp(j\theta_{HV}) \\ |S_{VH}|\exp(j\theta_{VH}) & |S_{VV}|\exp(j\theta_{VV}) \end{bmatrix} \cdot \begin{bmatrix} E_H \\ E_V \end{bmatrix}_{incident}$$

$$(3)$$

By comparing the lhs and rhs of Equation (3), we see that the scattering matrix, [S], is in the form of a transformation from an incident polarization to a scattered polarization. A unique 1:1 mapping of the scattering matrix [S] from the above linear polarization to any arbitrary orthogonal basis, such as circular, exists (see Emmons, 1983). Simple transformations between polarizations make the polarization scattering matrix a versatile description of the scattering behavior of many radar targets. Wiesbeck and Riegge (1987) describe a RCS measurement facility with high sensitivity, high dynamic range, and high accuracy. They describe the systematic and statistical errors involved with RCS measurements and a method of measuring these errors in terms of 16 error coefficients. All these coefficients can be determined by 4 independent calibration measurements of an empty room, and three different calibration targets.

The measured polarization scattering matrix [S] actually represents the product of a series of transformations that a radar wave must undergo (Macikunas and Haykin, 1986). First, the polarization characteristics of the transmitting antenna are encountered, then the propagation medium, then the target and clutter–producing environment, followed again by the propagation medium and receiving antenna. This product of polarization transformations can be written as,

$$[S] = [T]_{TxAnt}[T]_{Medium} \left([S]_{Target} + [S]_{Clutter}\right) [T]_{Medium}[T]_{RxAnt}$$

$$(4)$$

where $[T]$ is a 2×2 complex transformation matrix.

The equivalent transformation $[T]$, of the transmitting and receiving antennas indicate primarily the degree of cross–polar leakage, i.e., $\underline{S}_{HH}$ leaking into $\underline{S}_{HV}$ or $\underline{S}_{VH}$ or, equivalently, the like polar orthogonality, i.e. $\underline{S}_{VV}$'s independence of $\underline{S}_{HH}$. For a single dual–orthogonally–polarized antenna radar (monostatic), the matrices $[T]_{TxAnt}$ and $[T]_{RxAnt}$ can be combined into $[T]_{Ant}$. The corruption of the measured polarization scattering matrix due to $[T]_{Ant}$ can ideally be cancelled, for example, by finding $[T]_{Ant}^{-1}$ and premultiplying the measured scattering matrix. Estimation and cancellation of these antenna–related effects will be detailed later.

The propagation medium can also have a significant impact on the measured scattering matrix. For example, Faraday rotation can influence the polarization vector. This rotation will cause detectable non–symmetry in the received scattering matrix $[S]$, which can be easily cancelled, see Bickel (1965). Atmospheric precipitation is usually spherically–symmetric, leading to the generalization that the propagation medium is also linear, thus the reciprocity theorem can be invoked (Cloude, 1983), i.e., $|S_{HV}| = |S_{VH}|$, leading to a simplified matrix called the relative phase scattering matrix, $[S]_{rel}$. Under these conditions, the transformation matrix of the medium is close to the identity matrix, $[T]_{medium} = \begin{bmatrix} 1 & 0 \\ 0 & 1 \end{bmatrix}$. Note that in the relation

$$[S]_{rel} = \begin{bmatrix} |S_{HH}|\exp(j\theta_{HH}-j\theta_{HV}) & |S_{HV}| \\ |S_{HV}| & |S_{VV}|\exp(j\theta_{VV}-j\theta_{HV}) \end{bmatrix} \qquad (5)$$

the magnitude of the diagonal (cross–polar) responses have been made equal and their phase has been made zero. The phases of the two like–polarized terms have been defined in terms of this reference phase. The use of a reference phase (Emmons, 1983) stems from the fact that the absolute phase terms in Eqn. (2) are related to target range, and are thus not target–dependent parameters. Specification of the relative scattering matrix requires only three amplitudes and two phases as opposed to four amplitudes and four phases for the absolute phase scattering matrix as in Eqn. (2). Almost all polarimetric processing techniques are based on the less general relative scattering matrix as in Eqn. (5); the use of this equation can thus be assumed, unless the absolute phase scattering matrix is specified explicitly.

The polarization scattering matrix of the target, $[S]_{Target}$, is ideally the only remaining component. If this were true, as in the detection of airborne targets "in the clear", target detection, identification, and classification could be accomplished by comparing $[S]_{Target}$ to the target's known polarization signature. Various techniques to make this comparison will be presented later. Targets often are not "in the clear" and the return from the target resolution cell is corrupted by some form of clutter, thereby resulting in the ($[S]_{Target}$ +

$[\underline{S}]_{Clutter}$) term in Eqn. (4).   It is the polarimetric processor's task to distinguish the polarization characteristics of a target in clutter from the return from clutter only (no target).   When the clutter is dominant this becomes a difficult task, making more powerful polarimetric processing necessary such as vector filtering, polarization adaptation, pattern recognition, etc.   Many advanced techniques actually measure or use known polarization properties of the clutter to eliminate or minimize its effects.   This subject will be treated in greater detail later in Section 11.3.

11.2.2 THE POINCARÉ SPHERE

Since the state of polarization of a wave is comprised of several dimensions, it is necessary to use some form of projection or mapping to 2 dimensions, so that it can be easily visualized.

Déschamps (1951) describes the adaptation of an optics technique (Poincaré, 1892) in which every possible polarization state is mapped on the surface of a sphere.   It is easiest to explain the co–ordinates of this "Poincaré sphere" by first considering the definition of the most general polarization state, i.e., elliptical as in Figure 1.

Here, the geometrical parameters are: inclination angle θ and ellipticity angle τ.   For linear polarizations, the ellipticity angle is zero, and inclination angle describes the polarized wave's electric field direction.   Circular polarizations have ellipticity angle of $+45^0$ for LHCP and -45^0 for RHCP. Elliptical polarizations have values of ellipticity angle not equal to either

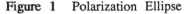

τ = the ellipticity angle θ = inclination angle

$-45^0 < \tau < 45^0$ $0 \leq \theta < 180^0$

$\tan \tau = \dfrac{b}{a}$ E = amplitude
                                        $= a^2 + b^2$

$\tau = \quad 0^0$ linear polar's

$\tau = -45^0$ RHCP

$\tau = \quad 45^0$ LHCP

Figure 1   Polarization Ellipse

$\pm$ 45⁰ or 0⁰.   These parameters are illustrated on the Poincaré sphere in Figure 2, along with the locations of some other commonly used polarizations.

11.2.3 A UNITARY POLARIZATION TRANSFORM

A unitary polarization transform [T], [S'] = $[T]^T$[S][T], will now be described (^{T}—denotes transpose).   This transform is unitary in that both the original and transformed basis vector pairs in Eqn. (6) (Cloude, 1983) or scattering matrices in Eqn. (7) (Huynen, 1982) have equal energy:

$$\underline{X}' \;=\; [T]\,\underline{X} \;=\; \frac{1}{\sqrt{1+\underline{\rho}^*\,\underline{\rho}}}\begin{bmatrix} 1 & -\underline{\rho}^* \\ \underline{\rho} & 1 \end{bmatrix}\underline{X} \tag{6}$$

where

$$\rho \;=\; \tan\,(\alpha e^{\tau\delta})$$
$$\overline{\underline{X}'} \;=\; \text{transformed basis vector}$$
$$\underline{X} \;=\; \text{original basis vector}$$

$$[S'(A'B')] \;=\; [T]^T[S(AB)][T] \tag{7}$$

$$[S(AB)] \;=\; [T]^*[S(A'B')][T]^{*T}\;.$$

In Eqn. (6) tanα is the magnitude of the polarization ratio, and δ is the phase.   These new parameters (α,δ) can be related to the Poincaré sphere/elliptical—polarization parameters (θ,τ) as follows:

$$\tan 2\theta \;=\; \tan 2\alpha\;\cos\delta \tag{8}$$
$$\sin 2\tau \;=\; \sin 2\alpha\;\sin\delta$$

Now polarization, $\hat{P}$, can be plotted using the above new parameters on a "Deschamps sphere"; see Figure 3.   These extra parameters give an alternate interpretation to the Poincaré sphere representation given in Figure 2.   This

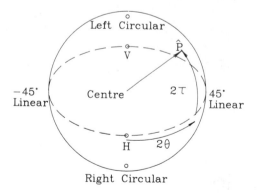

Note that the equator is the locus of all linear polarizations

Figure 2 The Poincaré Sphere.

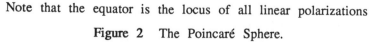

unitary transform will prove very useful when certain characteristic polarizations of targets are calculated in Section 11.3.1.

It would be convenient to project the 3–dimensional data of the Poincaré sphere to a 2–dimensional form. Various projections have been utilized in the past, including the modified polarization chart (Poelman, 1980). This chart is an equatorial projection of the Poincaré sphere that maps the sphere into two circular charts. Both charts have linear polarization represented along the circumference of the chart and circular polarization at the centers. Specifically, vertical and horizontal linear polarizations are located at the 9 and 3 O'clock positions, respectively, and ± 45⁰ linear polarizations occur at 12 and 6 O'clock on both charts. One chart differs from the other in that the circular polarization at the centre is LHCP in one chart and RHCP in the other.

More recently, Okishi (1986) has proposed a planar chart that has some practical advantages. This chart is obtained by a stereoscopic projection of the Poincaré sphere from the north pole to a plane touching the opposite pole. One of this chart's main advantages is that a circle on the surface at the Poincaré sphere will map to a circle in this new chart. This similarity could be quite important if one wishes to inspect the cluster shape of some polarization phenomenon on a 2–dimensional chart. The main disadvantage of the "Okishi" chart is that one (of two) circular polarizations will map to infinity on the chart. However, this last disadvantage may not prove to be a problem, depending on the application.

11.2.4 THE COMPLETELY–POLARIZED WAVE

By definition a completely–polarized wave has polarization characteristics that are time–invariant (Poelman, 1981). Conversely, non–polarized or partially–polarized waves have a polarization state that is time–varying. These two cases can be visualized as points on the Poincaré sphere that are either stationary over time or in motion, respectively. Generally, backscattered fields from moving objects, clutter, as well as interference fields will have large non–polarized components, while reflections from fixed large objects will be highly–polarized, i.e., small non–polarized components. The degree of polarization, p, of the wave is a measure of the ratio of the average polarized power to the total average power of the wave

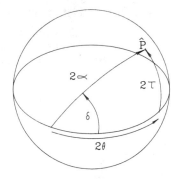

Figure 3 The Déschamps Sphere.

(Poelman, 1981), where $0 \le p \le 1$. For completely–polarized waves $p = 1$, and for completely unpolarized waves $p = 0$. It is important to note that the two outputs from orthogonally–polarized antennas will be equal for completely–unpolarized waves. Under such conditions, a radar system will have no polarimetric processing potential (Poelman, 1981).

Completely–polarized waves can be expressed as the complex addition of two orthogonal polarization basis vectors, such as vertical and horizontal linear polarization, as shown by

$$ h = \hat{P} = \underline{h}_A h_A + \underline{h}_B h_B \tag{9} $$

where h is the vector representing the time invariant polarization of the wave, (h_A, h_B) constitutes a pair of mutually orthogonal polarization vectors, and $\underline{h}_A$, $\underline{h}_B$ are the complex coefficients determining h; see Cloude (1983). From the coefficients $\underline{h}_A$ and $\underline{h}_B$ the complex polarization ratio, ρ, can be computed as in Equation (10) given below. The magnitude of $\overline{\rho}$ determines the orientation, and the phase of $\overline{\rho}$ determines the ellipticity, respectively, of the polarization ellipse given in Figure 1:

$$ \underline{\rho} = \frac{\underline{h}_B}{\underline{h}_A} = |\rho| \, e^{j\phi_\rho} \tag{10} $$

11.2.5   THE PARTIALLY–POLARIZED WAVE: STOKES VECTORS

The radar return is seldom completely–polarized; often it is partially–polarized, i.e., it is composed of both completely–polarized and fluctuating–polarization components. For such waves, the expressions in Equations (9) and (10) cannot completely describe the polarization phenomena. A more general approach was developed in the optics field by Stokes (1852), known as the Stokes vector approach. The Stokes vector is composed of 4 Stokes parameters which can be defined in terms of the time averages of the complex coefficients $\underline{h}_A$ and $\underline{h}_B$ (from Equation (9)) of the received wave. Note that since time averages and the relative phase between $\underline{h}_A$ and $\underline{h}_B$ are utilized, the Stokes vector can be used with non–coherent radar systems. Also, some difficulties may arise if the illuminating polarizations fall along the targets null polarizations (Mieras, 1983). Nevertheless, we may write the Stokes vector as

$$ \overline{g} \equiv (g_0, g_1, g_2, g_3) \tag{11} $$

where

$$ \overline{g}_0 = <|\underline{h}_A|^2> + <|\underline{h}_B|^2> $$

$$ \overline{g}_1 = <|\underline{h}_A|^2> \cdot <|\underline{h}_B|^2> $$

$$\bar{g}_2 = \; < 2|\underline{h}_A| \cdot |\underline{h}_B| \; \cos\phi_{B-A} >$$

$$\bar{g}_3 = \; < 2|\underline{h}_A| \cdot |\underline{h}_B| \; \sin\phi_{B-A} >$$

and $< \cdot >$, $\bar{}$ denote time averages.

The various parameters in Equation (11) have physical significance: g_0 is the incident power density, g_1 is the tendency to resemble horizontal or vertical polarization, g_2 is the tendency toward $\pm$ 45^0 linear polarization, and lastly g_3 is the tendency of the wave to be left- or right-hand circular polarized; see Table 1 for some common Stokes vectors along with values of the parameter ρ (Cloude, 1983).

Table 1   Values of ρ For Some Common Polarizations

Polarization	ρ	Stokes Vector
Horizontal	0	(1 1 0 0)
Vertical	∞	(1 −1 0 0)
+ 45^0 Linear	1	(1 0 1 0)
Left Circular	j	(1 0 0 1)
Right Circular	$-j$	(1 0 0 −1)

The Stokes parameters are not independent if the wave is completely polarized (coherent), as shown by

$$g_0^2 = g_1^2 + g_2^2 + g_3^2 = \sum_{i=1}^{3} g_i^2 . \qquad (12)$$

If the power, g_0, is taken as the radius of a sphere, the three remaining parameters will define a Cartesian polarization space equivalent to the Poincaré sphere representation presented in Figure 2.   In this figure, we have

$$
\begin{aligned}
g_0 &= 1 \\
g_1 &= \cos 2\tau \; \cos 2\theta \\
g_2 &= \cos 2\tau \; \sin 2\theta \\
g_3 &= \sin 2\tau \qquad . \qquad (13)
\end{aligned}
$$

This equivalence gives a convenient method of transforming the three easily measurable parameters g_1, g_2, and g_3 to the geometrical representation of the Poincaré sphere.

If the Stokes parameters are used to represent transmitted and received polarization parameters rather than dual–orthogonal polarizations, a different form of scattering matrix results. The resulting 4×4 matrix with 16 real elements is often called the Stokes reflection or Mueller scattering matrix:

$$[M] = \begin{bmatrix} m_{11} & m_{12} & m_{13} & m_{14} \\ m_{21} & m_{22} & m_{23} & m_{24} \\ m_{31} & m_{32} & m_{33} & m_{34} \\ m_{41} & m_{42} & m_{43} & m_{44} \end{bmatrix} \tag{14}$$

It can be used to describe the input/output relationship of an antenna, target, environment, etc. We may thus write

$$g_{out} = [M]\, g_{in} \tag{15}$$

Both the scattering matrix and Mueller scattering matrix are useful in radar polarimetry. The two matrices can be related in a straightforward manner providing that first, the radar is monostatic, and second, that a modified set of Stokes parameters are utilized, as shown by:

$$\begin{aligned} g_m &= g_m(\, I_A,\ I_B,\ U,\ V\,) \\ g_{m0} &= |\underline{h}_A|^2 = I_A \\ g_{m1} &= |\underline{h}_B|^2 = I_B \\ g_{m2} &= 2\mathrm{Re}\{\ \underline{h}_A\ \underline{h}_B^* \ \} = U \\ g_{m3} &= -2\mathrm{Im}\{\ \underline{h}_A\ \underline{h}_B^* \ \} = V \quad . \end{aligned} \tag{16}$$

Using the new Stokes parameters, [M] is expressed in terms of the conventional scattering parameters as:

$$[M_{mono}] = \begin{bmatrix} |S_{AA}|^2 & |S_{AB}|^2 & \mathrm{Re}(S_{AA}S_{AB}^*) & \mathrm{Im}(S_{AA}S_{AB}^*) \\ |S_{BA}|^2 & |S_{BB}|^2 & \mathrm{Re}(S_{BA}S_{BB}^*) & \mathrm{Im}(S_{BA}S_{BB}^*) \\ 2\mathrm{Re}(S_{AA}S_{BA}^*) & 2\mathrm{Re}(S_{AB}S_{BB}^*) & \mathrm{Re}(S_{AA}S_{BB}^*+S_{AB}S_{BA}^*) & \mathrm{Im}(S_{AA}S_{BB}^*-S_{AB}S_{BA}^*) \\ -2\mathrm{Im}(S_{AA}S_{BA}^*) & -2\mathrm{Im}(S_{AB}S_{BB}^*) & -\mathrm{Im}(S_{AA}S_{BB}^*+S_{AB}S_{BA}^*) & \mathrm{Re}(S_{AA}S_{BB}^*-S_{AB}S_{BA}^*) \end{bmatrix} \tag{17}$$

The conventional scattering matrix parameters are defined in terms of Stokes parameters as follows:

$$|S_{AA}| = M_{11}^{1/2}, \qquad \phi_{AA} = \phi_{AB} + \tan^{-1}(M_{14}/M_{13})$$

$$|S_{AB}| = M_{12}^{1/2}, \qquad \phi_{AB} \text{ is arbitrary}$$

$$|S_{BA}| = M_{21}^{1/2}, \qquad \phi_{BA} = \phi_{AB} + \tan^{-1}\left[\frac{M_{34} + M_{43}}{M_{33} - M_{44}}\right]$$

$$|S_{BB}| = M_{22}^{1/2} \qquad \phi_{BB} = \phi_{AB} + \tan^{-1}(M_{42}/M_{32}) \tag{18}$$

In the monostatic case, $|S_{AB}| = |S_{BA}|$ and $\phi_{AB} = \phi_{BA}$, resulting in further simplifications of Eqn. (18). It is also interesting to note that matrix $[M_m]$ has only nine independent elements. The above transformation is given by Boerner (1981), and similar transformations can be found in Huynen's Ph.D. thesis (1970).

11.2.6   THE COHERENCY MATRIX

Wiener showed that the polarization state of a wave can also be completely characterized by a 2×2 matrix which he named the coherency matrix (Wiener, 1927, 1930). This matrix can be measured using a pair of mutually—orthogonal polarization amplitudes at right angles to the direction of propagation, A and B (as used in Equation (9)). The coherency matrix is defined by

$$[J] = \text{coherency matrix} = \begin{bmatrix} J_{11} & J_{12} \\ J_{21} & J_{22} \end{bmatrix}$$

$$[J] = \begin{bmatrix} \langle AA^* \rangle & \langle AB^* \rangle \\ \langle A^*B \rangle & \langle BB^* \rangle \end{bmatrix} \tag{19}$$

Lately, this alternate representation of the polarization state of a wave has received renewed interest, possibly due to its ability to deal with partially—polarized waves.

The coherency matrix has a number of convenient properties; see Ko (1962) and Nespor, et al. (1984). First, the backscattered power can be easily expressed as:

$$\text{Tr}([J]) = J_{11} + J_{22} \tag{20}$$

where $\text{Tr}(\cdot)$ denotes the trace of the matrix. $\text{Tr}([J])$ is not polarization–invariant, but $\text{Tr}([J]+[J]_{\perp})$ is, see Stock and Trogus (1984). Also, this invariance is equivalent to the following invariance (Graves, 1956):

$$\text{Tr}([S]^H[S]) \quad = \quad \text{Tr}([J]+[J]_{\perp}) \tag{21}$$

where [S] is the scattering matrix from Eqn. (3), $\cdot^H$ denotes the Hermitian transpose, and $[J]_{\perp}$ is the coherency matrix for the "orthogonal" target, see Stock and Trogus (1984). The determinant of matrix $[J]$ is always non–negative;

$$\det([J]) \quad = \quad J_{11}J_{22} - |J_{12}|^2 \quad \geq \quad 0 \tag{22}$$

It is also obvious from Eqn. (19) that $J_{12} = J_{21}^*$, since $|S_{12}| = |S_{21}|$ by reciprocity. The complex correlation between the two orthogonally–polarized channels is given by

$$\underline{\mu} \quad = \quad |\underline{\mu}|e^{j\beta} \quad = \quad \frac{J_{12}}{\sqrt{(J_{11}\,J_{22})}} \tag{23}$$

The special case of completely polarized waves corresponds to $\det([J]) = 0$ in Equation (22), while completely unpolarized waves have the cross terms $J_{12} = J_{21} = 0$, and the diagonal terms $J_{11} = J_{22}$. The degree of polarization, p, can also be easily expressed in terms of the coherency matrix parameters as

$$p \quad = \quad \sqrt{1 - \frac{4\det([J])}{(J_{11} + J_{22})^2}} \quad . \tag{24}$$

The Stokes parameters in Equation (11) also bear close resemblance to the elements of the coherency matrix, as shown by

$$\bar{g}_0 \quad = \quad J_{11} + J_{22}$$
$$\bar{g}_1 \quad = \quad J_{11} - J_{21}$$
$$\bar{g}_2 \quad = \quad J_{12} + J_{21}$$
$$\bar{g}_3 \quad = \quad j(J_{12} - J_{21}) \tag{25}$$

11.2.7   THE 4 x 4 HERMITIAN COHERENCY MATRIX

Cloude (1986a, 1986b) describes a homomorphism between the Mueller matrix and a real 6–dimensional target space. This homomorphism is similar to the algebraic technique used to relate the complex polarization state given by time–invariant polarization $\underline{h}$ to Poincaré sphere coordinates. The practical utilization of this technique will be described below.

To begin with, the Mueller (4×4 real) matrix is measured. This is accomplished by measuring the (time–averaged) Stokes vector $\bar{g}$ (four parameters) of a target under 4 different illuminating polarizations, say linear 0^0, 45^0, 90^0, and left–hand circular. These 16 measurements are used to solve for the 16 unknown Mueller matrix elements using equations. Conventional processing would make use of this "raw" Mueller matrix. The Mueller matrix [M] is then used to compute the 4×4 Hermitian target coherency matrix $[T_c]$ (see Cloude (1986a)). The target polarization state can then be uniquely mapped onto the surface of a 6–dimensional hypersphere (analogous to using the Poincaré sphere mapping in 3 dimensions). More importantly, one can calculate the eigenvectors of $[T_c]$. Using this eigen–analysis, the target coherency matrix $[T_c]$ can be decomposed as an eigenvalue–weighted sum of 4 target components. Cloude (1986b) states that his decomposition should be capable of extracting scattering matrix data in the presence of additive receiver noise, and that this eigenvector decomposition is the only such fundamental decomposition of partial targets described in the literature.

11.3   TARGET / CLUTTER DISCRIMINANTS

11.3.1 OPTIMAL POLARIZATIONS: THE EIGENVALUE PROBLEM

The use of optimal polarizations (like– and cross–polar nulls) as radar discriminants seems to be a very promising and yet an under–investigated area. Several orientation–invariant target or clutter descriptors are based on the concept of one pair of cross–polarized nulls and one pair of like–polarized nulls representing transmitted polarization vectors for which the received cross–polarized and like–polarized components are zero, respectively. These null polarizations can be easily found through an eigenvalue analysis of the scattering matrix (5). In fact, eigenvalue analysis also gives the solution to the unitary transform $[\underline{T}]$, which diagonalizes the scattering matrix.

The solution of the matrix cross–polar nulls (sometimes called maximum polarizations) is the solution to the relation:

$$[S(AB)]\ \underline{x}\ =\ S\ \underline{x}^{*} \tag{26}$$

The complex conjugation, denoted by an asterisk, arises due to difference in transmit and receive polarization vector orientations. Eigenvectors, $\underline{x}$, will be orthogonal to each other due to algebraic considerations, lying diametrically opposite to one another on the Poincaré sphere. The like–polar nulls can be found by a solution of the unitary transform $[\underline{T}]$, which diagonalizes $[\underline{S}]$ to the form:

$$\begin{bmatrix} S_1 & 0 \\ 0 & S_2 \end{bmatrix} = \begin{bmatrix} m\ e^{i\delta} & 0 \\ 0 & m\ \tan^2\gamma\ e^{-i\delta} \end{bmatrix} \tag{27}$$

where $m = |S_1|$, and $\gamma - 90^0 \quad \alpha$.

Two possible solutions are given for

$$\rho^2 = -\frac{S_1}{S_2} = \tan^2\alpha \; e^{j2\delta}$$

by,
$$S_1 = \tan\alpha \; e^{\tau\delta}$$

$$S_2 = -\tan\alpha \; e^{\tau\delta} \tag{28}$$

These two pairs of null polarization solutions in Eqn. (28) can be plotted on the Poincaré sphere to create the so–called Huynen fork, as in Figure 4; see Huynen (1970 or 1982) for more details.

An alternative calculation of the actual co–ordinates of the cross–polarized and like–polarized nulls can be found by solving scattering matrix [S] for the parameter ρ from the unitary transformation (see Daley (1983) and Boerner (1981)), as shown by:

$$0 = S_{11} + \rho^2 S_{22} + \rho(S_{12} + S_{21}) \tag{29}$$

$$0 = -\rho^* S_{11} + \rho \; S_{22} + S_{12} - \rho\rho^* S_{21} \tag{30}$$

$$r = P = |S_{11}|^2 + |S_{22}|^2 + 2|S_{12}|^2 \tag{31}$$

For the like–polar nulls, use Eqn. (29) to solve for ρ, use Eqn. (30) to find ρ for the cross–polar nulls, and use Eqn. (31) to solve for r, the radius of

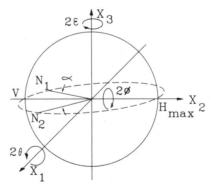

N$_1$ N$_2$ — coplanar nulls
θ — related to target rotation
ϕ — skip angle, related to the number of bounces
ε — helicity angle, related to target symmetry
α — fork angle

Figure 4   Huynen Fork.

the Poincaré sphere. The conventional polar co—ordinates of the null polarizations can now be found using the following relations with appropriate $\underline{\rho}$ for either the like— or cross—polar nulls:

$$q = \frac{1 - j\underline{\rho}}{1 + j\underline{\rho}}$$

$$\tan\phi' = \frac{-\mathrm{Im}(q)}{\mathrm{Re}(q)}$$

$$\cos\theta' = \frac{-1 + |q|^2}{1 + |q|^2} \tag{32}$$

where, r, θ', and ϕ' are the conventional spherical co—ordinates of the optimal polarization.

Yet another alternate specification of the optimal polarizations is given by Nespor, et al. (1984) in terms of the coherency matrix parameters. In fact, one cross—polar null polarization, one like—polar null polarization along with an amplitude factor (equivalent to polarization sphere radius) will completely describe the relative scattering matrix. These null parameters can then be considered as an alternative method of specifying the five independent scattering matrix parameters outlined in Equation (5). Additionally, the values of the cross and/or like—polar nulls can be used directly as target discriminants because they are polarization base—independent and the nulls relate directly to target symmetry and its rotation about the line of sight.

A simulation study of the optimal null polarizations of various complex targets is described by Raven (1985). In this study the co—polarized null polarizations (N_1 and N_2 from Figure 4) are calculated for targets consisting of combinations of conducting thin wires, dihedrals and trihedrals with random amplitudes, phases and orientations as well as single rotating composite dihedrals. The composite dihedrals are dihedral corner reflectors with the two faces composed of two diverse materials such as steel and seawater, steel and dry ground, wood and moist ground, etc. The clustering of the null polarizations is displayed on an equal area projection of the Poincaré sphere in 2 dimensions. Every corresponding pair of null polarizations are joined by a line in this projection. Dihedral reflectors, singly or in combination, exhibit easily identifiable horizontal joining lines between pairs of co—polar nulls. If the SNR is decreased (below 10 dB) or if trihedral targets are added, the above predictable line structure disappears. Raven also demonstrates a considerable migration or movement of pairs of co—polarized nulls for the composite dihedrals described above as the angle of incidence is varied.

A recent reexamination of the foundations of radar polarimetry introduces a new solution to the optimal polarization problem (Kostinski and Boerner, 1986). In this study, a solution to the optimum receiver voltage problem is presented in terms of the theory of Hermitian forms. Traditional optimization requires diagonalizing the scattering matrix using a pseudo—eigenvalue technique. The new technique has the advantages of providing a clearer physical interpretation, avoiding the diagonalization of [S], and of treating all forms (symmetrical, asymmetrical, monostatic and bistatic) of the scattering matrix in a similar fashion. In addition to the above contribution, Kostinski and Boerner also restate such fundamentals as choice of coordinate systems and the basic equations of polarimetry. The investigation does, however, uncover an inconsistency in that two forms of the scattering matrix are needed depending on whether [S] appears in the voltage equation or in the polarization state equation. A satisfactory solution to this dual formalism has not yet been found.

The new solution to optimum polarization presented by Kostinski and Boerner (1986) is extended to maximize the ratio of echoes from two received targets (Kostinski and Boerner, 1987). The new technique called polarimetric contrast optimization specifies the polarization of the antenna required to maximize the ratio between the received echoes from two different target classes. The improvement here is that (again, using Hermitian forms) the amount of computation required to optimize the contrast is reduced.

11.3.1.1 DISCRIMINANTS BASED ON OPTIMAL POLARIZATIONS

Daley (1981, 1983) has studied the loci of optimal polarizations produced by a simulated 50' vessel in simulated sea clutter and noise. The loci of points were plotted in a θ, ϕ-space, where θ, ϕ are the conventional spherical polar co-ordinates of the optimal polarizations. Mieras (1983) investigated the like-polar null loci produced by two simple targets, a rotating dumbbell, and a cylinder, both with various dimensions. The null loci for this study is plotted on a 2-dimensional equal-area projection of the Poincaré sphere surface. The study of Mieras did not address the question of how to use the loci-plots as discriminants, but did offer some interesting discussions on the variance of loci points and on the tendency of the simple targets to generate null loci typical of objects that were very different physically. This last phenomenon may in fact be irrelevant if the RCS of these simple objects is quite small at the aspect angles which generated the confusing nulls. Lastly, Boerner (1981) gives numerical data describing the expected scattering matrix and like-polar and cross-polar nulls for sea clutter, and Nespor, et al. (1984) describe the variance of like-polar nulls for various hydrometeors.

Two potential target discriminants based on the like- and cross-polar nulls have been investigated by Daley (1981, 1983). The first discriminant is the spherical variance of the like- and cross-polar nulls, as shown by

$$ S \;=\; 1 \,-\, \frac{R}{N} \tag{33} $$

where

$$N \;=\; \text{number of sample points}$$
$$S \;=\; \text{spherical variance of like- or cross-polar null} \quad,$$

and

$$R \;=\; \text{spherical mean direction} \quad.$$

The spherical mean direction itself is defined by

$$R^2 = (\Sigma_i k_i)^2 + (\Sigma m_i)^2 + (\Sigma n_i)^2 \qquad (34)$$

where k, m, and n are the direction cosines from the centre of the sphere to a null point:

$$
\begin{aligned}
k &= \cos\theta \, \cos\phi \\
m &= \cos\theta \, \sin\phi \\
n &= \sin\theta \quad .
\end{aligned}
$$

The second target discriminant is the circular variance of the longitudinal component of the nulls, given by

$$C = 1 - \frac{R}{N} \qquad (35)$$

where

$$
\begin{aligned}
C &= \text{circular variance of the nulls} \\
N &= \text{number of sample points} \\
R &= \text{circular mean direction} \quad ,
\end{aligned}
$$

and

$$
\begin{aligned}
R^2 &= (\Sigma_i k_i)^2 + (\Sigma m_i)^2 \\
k &= \cos\phi \\
m &= \sin\phi \quad .
\end{aligned} \qquad (36)
$$

Daley (1981) reports that the above variance discriminants could provide a detection performance better than horizontal amplitude only (a conventional discriminant). The choice of a discriminant was based on the arbitrary criterion that the discriminant should have higher S/N than horizontal amplitude for more than 50% of the grazing angles tested. In general, the variances tend to decrease with increasing target size, or equivalently, target–to–clutter and noise ratio. For large targets the circular variance of the cross–polarized nulls appears to be the best discriminant over most grazing angles. Small targets, (T/C $\simeq$ 0 dB), are easily detected at large grazing angles using any of the spherical or circular variance discriminants, but these discriminants become inferior to horizontal amplitude at small grazing angles (noise dominated return). These variances appear to be promising discriminants, provided that noise is not allowed to dominate the return, as is the case at very low grazing angles.

Another very promising discriminant appears to be the natural logarithm of the transformation parameter, ρ, used in Section 11.3.1. The use of $\ln(\rho)$ seems advantageous when attempting to distinguish polarization–rotating targets (two–bounce) such as the dihedral from ordinary non–depolarizing objects. The results of Daley (1983) clearly show $\ln(\rho)$ as being roughly proportional to ship size for a wide range of grazing angles, and very close to zero for a dihedral reflector (RCS = 30 dBm^2) for a similar range of grazing angles.

11.3.1.2   OPTIMAL POLARIZATIONS IN REMOTE SENSING

Airborne coherent polarimetric radar data has been collected by a group at Jet Propulsion Laboratory (Zebker, et al., 1987). Using a high–resolution synthetic–aperture radar (SAR) with a dual–polarized antenna, Zebker, et al. have measured the scattering matrix of various forms of ground clutter, i.e., forests, urban areas, and sea clutter.

Based on measurements of amplitudes and phases of backscattered horizontal and vertical transmit signals using horizontal or vertical receive antennas, any arbitrary pair of transmit and receive polarizations can be synthesized. The SAR images of land and sea clutter regions are shown to vary greatly in the relative contrasts between different image features with different transmit/receive polarizations.

Next, using a polarization chart consisting of a 2–dimensional surface where the two longitudinal components are the Poincaré sphere orientation angle, θ, and ellipticity angle, τ, and where the height of the surface is the relative intensity, it is easy to identify the optimum polarizations of the radar echoes. Alternatively, the ratio of polarization signatures can be computed between two features that are to be distinguished. By using the maximum polarization indicated by this ratio to process the SAR image, the relative contrast between the two features may be increased. This scheme was effectively demonstrated by Zebker, et al., to enhance visibility of ships in sea clutter, the contrast of lava flows and to enhance the contrast between forest and clear–cut field areas.

The above data was analyzed in a different manner by Van Zyl (1987). In Van Zyl's approach, the received data from the polarimetric SAR is converted to its Stokes vector form. Utilizing the data in this form, the polarized and non–polarized nature of the scattering is preserved in the measurements. Van Zyl's departure from conventional processing is his calculation of the optimal polarizations using the Stokes scattering operator (Van Zyl, 1987). He shows that for non–polarized scatterers, up to 6 optimum polarizations can exist where only 4 were possible using the 2×2 scattering matrix approach (see Section 11.3.1). Additionally, the technique can still be used on completely polarized data, in which case only 4 optimal polarizations exist.

11.3.2   COVARIANCE–MATRIX–BASED STATISTICS

The covariance between the various elements of the relative scattering matrix in Equation (3) can be used to form the covariance matrix (Daley, 1981). This new matrix is of dimensions 3×3, since only 3 complex terms of the scattering matrix were taken as unique in Equation (5). For sample averaging of N samples, the sample (integrated) covariance matrix is given by

$$[\underline{C}] \;=\; \begin{bmatrix} \underline{C}_{11} & \underline{C}_{12} & \underline{C}_{13} \\ \underline{C}_{21} & \underline{C}_{22} & \underline{C}_{23} \\ \underline{C}_{31} & \underline{C}_{32} & \underline{C}_{33} \end{bmatrix} \tag{37}$$

where

$$\underline{C}_{ij} = \frac{1}{N} \sum_{m=1}^{N} Z_{im} Z_{jm}^{*}$$

and

$$Z_1 = |S_{11}| \exp(j\theta_{11} - j\theta_{12})$$

$$Z_2 = |S_{22}| \exp(j\theta_{22} - j\theta_{12})$$

$$Z_3 = |S_{12}| \exp(j\phi^0)$$

Usually the parameters, Z_i, are normalized as

$$S_{ij} \text{ (normalized)} = S_{ij} / \sqrt{P} \tag{38}$$

where P is the total power of the received echo.

One detection statistic based on the normalized covariance matrix is the matrix difference measure (Daley, 1981):

$$\delta = \sum_{i=1}^{N} \sum_{j=1}^{N} |A_{ij} - B_{ij}|^2 \tag{39}$$

where, $[A]_{ij}$ is the reference matrix (clutter plus noise), and $[B]_{ij}$ is the sample matrix including the target. Daley found that using δ as a detection statistic gave better performance than horizontal-amplitude only and better than circular variance of the like-polar nulls (described earlier in Section 11.3.1) for low target-to-clutter (T/C) ratios (T/C ratio $\simeq$ −20 to −10 dB) and worse for high T/C ratios. The matrix difference measure performed better than $\sqrt{\sigma_{HH}}$ for high depression angles and worse than $\sqrt{\sigma_{HH}}$ for low depression angles.

11.3.3 POLARIZATION VECTOR PROCESSING

Poelman (1981) proposed a technique whereby the effective transmit and receive polarizations of a radar system are adapted electronically through polarization-vector filtering. In essence, the polarization state of interference, clutter, or precipitation is estimated and then used to define a polarization vector filter. This vector filter is then applied to the received data, which is then subsequently submitted to the target detector. For this technique to be implemented, the radar must be dual-orthogonally polarized with dual-polarized transmitting capability, i.e., the same requirements for measuring the polarization scattering matrix. The adaptation technique capitalizes on the fact that targets are generally highly polarized, and interference fields, and backscatter from moving objects etc., are generally non-polarized. The desired adapted polarization for transmission is the one that minimizes the non-polarized reflected echo, while the adapted receive polarization enhances the echo of the target.

The above vector filter is estimated somewhat differently for clutter, and for interference. For interference, the interference vector is measured and averaged along a moving window, and similarly for homogeneous clutter, i.e., sea, flat land, etc. The performance for these cases is determined by the degree of polarization during the observation window. Typically, a 2–dB loss in echo level can be expected with a 10–dB improvement in target–to–clutter (T/C) ratio. For the dipole–cloud and aircraft model in which there is 85% polarization noise, simulated results by Poelman (1980, 1981) show 14 dB of suppression. Inhomogeneous clutter must be handled on a cell–by–cell basis in which clutter vectors from the previous scan must be stored and then recalled to produce the vector filter. In this situation, suppression is optimum when the clutter exhibits fixed–object characteristics. The vector filtering problem can be viewed as filtering two loci on a Poincaré sphere, one due to the target and the other due to interference. The filters prescribed for the polarization filtering are digital realizations of n'th order multinotch logic–product (MLP) filters (Poelman, 1980). This filter is composed of n polarization space notch filters with outputs fed to a logic product device which selects the smallest of all inputs for further processing (this last characteristic implies a non–linear filter). The performance of all such filters will improve as the spherical distance between the target and interference loci increases and their spherical variances decrease. Lastly, Poelman states that the filtering techniques depend on the application of fast adaptive algorithms to adapt quickly to the interference and/or clutter polarizations.

11.3.3.1 APPLICATION TO SURVEILLANCE RADAR

A simple scheme of improving the clutter suppression of surveillance radars employing circular polarizations has been described by Nathanson (1975). Such radars should in theory be capable of around 25 dB of rain clutter suppression, based on their antenna polarization purity; but they seldom achieve over 15 dB of suppression. The problems are in fact many–fold. First, the raindrops are not spherical but become elliptical in shape, second, the horizontal and vertical components of the wave do not suffer equal attenuation, and lastly reflections from the ground or water surface also become elliptically polarized. The proposed solution is an adaptive polarization canceller (APC) circuit having a similar structure to a homodyne modulator, where it is desired to cancel the residual precipitation echo in the like–polarized channel using the precipitation echo in the cross–polarized channel.

The above adaptive polarization canceller (APC) proposed by Nathanson (1975) has also been investigated by Giuli, et al. (1982). In this study, Giuli evaluated the performance of the APC by using simulated of the target data. The data was obtained from a model based on decomposition of the target into polarized– and non–polarized components closely approximating the echo from a real aircraft. An improvement of 9 dB target–to–clutter ratio was obtained using the APC on the above aircraft model echo in rain clutter (rain rate of 2.5 mm/hr).

A further improvement to the APC suitable for suppressing interference and jamming signals has been proposed by Giuli, et al (1982). This new canceller called the symmetrical adaptive polarization–canceller (SAPC) is composed of two APC's to perform cancellation on the two orthogonally polarized channels simultaneously. The output from this device is chosen from the canceller having the lowest disturbance signal.   This SAPC is shown by Giuli to provide a high degree of interference (jamming) suppression when the jammer is highly polarized and has significantly different polarization from the desired target.

Fossi, et al. (1984) show measured polarization responses from aircraft and ground clutter using coherent air traffic control (ATC) radar.   The output in the form of modified polarization charts demonstrates a high degree of fluctuation in scan–to–scan of aircraft polarization but little such fluctuation in ground clutter polarization.   Using Poelman's multinotch logic–product filter (1981), a ground clutter suppression of 14 dB was realized.   The output of the above filter can still be processed to extract doppler information.   Fossi, et al. used maximum–entropy method spectrum analysis on the short data sequences (16 samples) to estimate the doppler shift.   It was demonstrated that the doppler resolution was increased and spectral bias was reduced when the 2–channel Morf algorithm was used on the two receive channels compared to single receive channel spectrum estimation using the Burg algorithm.   The improvement in the spectral estimates using the dual polarized data can be explained by increased total available power using two channels and the fact that the received target signals are quite correlated and the clutter in the two receive channels are generally less correlated.

Giuli, et al. (1985) have suggested yet a further improvement in the area of adaptive polarization cancellers.   Their suggestion is to use an adaptive polarization canceller (APC) to find the minimum polarization estimate for the multinotch logic–product (MLP) filter.   The combined filter is more robust in that it has a wider suppression area while still tracking the minimum polarization.   The APC and MLP–APC were compared under strong jamming conditions (target–to–jamming T/J ratio $\simeq$ 0 dB) for an aircraft target.   The APC filter alone improved the T/J ratio by 14.5 dB while a 19–notch MLP–APC filter improved T/J ratio by 17.0 dB.   This new type adaptive polarization scheme possesses a substantial processing gain in the presence of jamming signals.

11.3.3.2 APPLICATION TO SEA / COASTAL RADAR

The use of adaptive polarization cancellers has not been limited to air traffic control (surveillance) radars.   Lammertse, et al. (1984) describe use of multipolarized coastal/harbour marine radars and adaptive polarization cancellers to suppress sea clutter echo.   The requirement is to find a method of measuring the polarization characteristics of spiky sea clutter so that the appropriate transmit/receive polarizations can be selected.   The goal for the multipolarized coastal/harbour radars is a reduction of sea clutter by up to 12 dB using adaptive polarization cancellers.

11.3.4 STOKES PARAMETERS AND MUELLER (STOKES) MATRIX

The Stokes vector or the Mueller (Stokes) matrix (described earlier in Section 11.2.5) is the most complete description of the polarization scattering process.   Consequently, it should provide many promising detection statistics. Recall from Section 11.2.5 that the time–averages of different quantities were required to define the Stokes parameters; consequently, the Mueller (Stokes) matrix is a time–averaged matrix as well, usually denoted by $[R] = <[M(t)]>$.

Measurement of the Mueller matrix for an ordinary conductive cylinder was undertaken by Mieras (1983). The results of the simulation demonstrate rapid fluctuation of different Stokes parameters with target aspect, but show an overall qualitative similarity when the Stokes parameters are obtained from different transmit polarization bases. Boerner (1981) gives a typical time–averaged Mueller matrix for sea clutter based on many different observations of the sea with polarimetric radar. The elements of the previous two Mueller matrices differ greatly because in the first case the target was stationary, and in the second it was in motion, i.e., time–varying.

In general, non time–varying targets can be fully described by five independent parameters. For time–varying targets nine parameters become independent (the above result for sea clutter was specified based on nine Mueller matrix parameters). Based on this, Huynen (1970, 1982) has developed several decomposition theorems to express the time–averaged Mueller matrix, [R], as the sum of a stationary (non time–varying) component, [M_o], and a noise component, [N], as shown by

$$[R] = [M_o] + [N] \tag{40}$$

Note that the matrix [R] has 9 independent parameters, and [M_o] (stationary component) has 5 independent parameters, as expected. The matrix [N] has the remaining 4 independent components, representing the time–varying components of the signal. Huynen proposes to use this last matrix (named the N–target) to classify different clutter and target types. But the decomposition required to find [M_o] and [N] presented by Huynen (1970) has been shown to have no strong physical basis by Cloude (1985). Furthermore, the Huynen decomposition was demonstrated as being one of an infinite number of such decompositions. Cloude (1985) derives a new decomposition theorem that is polarization–base invariant, utilizing the eigenvectors of the coherency matrix. Cloude then compares this new decompositioon to the decomposition proposed by Huynen.

Another study of interest is reported by Holm (1984). His approach to stationary target/clutter discrimination is as follows. The four Stokes parameters are measured and used to define a 4–dimensional feature space. Pattern recognition techniques are then applied to obtain decision surfaces separating targets from clutter and then to make a decision as to whether the echo is a target or clutter. As with Poincaré sphere discrimination, the performance of a pattern recognition technique depends upon the target/clutter separation in the chosen feature space. For high range or cross–range resolution radars, Holm prescribes the use of 4N or 4NM dimensional spaces, where N is the number of range samples including the target, and M is the number of cross–range samples including the target, so as to further increase the separation between the desired target and clutter clusters. For the case of moving (even slightly) complex targets, the target and clutter clusters in the feature space may overlap decreasing the separability of target and clutter. This overlap may be lessened through the use of time–averaged Mueller matrices or the decomposed matrix, [M_o], of Equation (40).

The study of Daley (1983) gives time–averaged Mueller matrices for two types of target. The first target is a large ship, and the second is a dihedral corner reflector of roughly equal RCS. A large difference in the

Mueller matrix M_{33} element is found to be evident: for ship, $M_{33} = -0.03$, and for dihedral, $M_{33} = -0.22$. This result may indicate the utility of using time–averaged Mueller matrices for the discrimination of co–operative and yet distinctive targets such as dihedrals from large man–made metal objects.

Vachula and Barnes (1983) view the detection of a target in clutter as a simple binary hypothesis testing problem. Their simulation utilizes a target composed of a combination of odd– and even–bounce reflectors illuminated by circular polarization in homogeneous random dipole clutter. In such clutter background the two received polarizations are decorrelated. Vachula and Barnes derive several Stokes–parameter–based detection statistics as well as the pdf's and joint conditional pdf's. For the given target/clutter model, the likelihood ratio (g_o+g_3) is found to have the best performance, but unlike the single–channel Marcum–Swerling detector it is not constant false–alarm rate (CFAR). Other possible detection statistics that are CFAR are worse than the Marcum–Swerling detector. But, Vachula and Barnes believe that all the detection performance should improve in non–homogeneous or spatially–varying clutter.

11.3.5 POLARIMETRY WITH NON–COHERENT RADAR

11.3.5.1 FEATURE DISCRIMINATION IN REMOTE SENSING

Multipolarized radar is capable of realizing greatly improved discrimination between different classes of targets. For example, conventional microwave remote sensing radars usually have only a single polarization capability. Using this single polarization limits the amount of information about the cause of the underlying scattering that can be deduced. Carver (1985) demonstrates greatly improved discrimination between different types of targets when multi–polarization capability is added to remote sensing radars. The reason for this improvement in discrimination is related to the manner in which microwaves interact with targets, different incident polarizations producing different echoes. Additionally, the cross–polarized channel also furnishes yet more information about the scattering process. A graphic display of the power of polarization was provided by presenting several radar images of the same area using either conventional single polarizations or using extra polarization information. Greatly enhanced discrimination between formerly indistinguishable target classes could be achieved using either a color coded image, where the echo levels of HH, VV, and VH polarizations were represented by level of colors red, green and blue, respectively, or by using the HH echo as the brightness parameter and the phase difference between HH and VV receive polarizations as the color. Note that the phase difference between received channels can be found using a coherent receiver with either a coherent or non–coherent radar transmitter.

11.3.5.2 AIRCRAFT DETECTION (VIDEO ADDITION)

Dual orthogonal received signals can be used to improve the detection/discrimination performance of non–coherent radars. Two techniques currently exist whereby the extra information in the second (cross–polarized) channel is utilized, namely, video addition and cross–to–like echo ratio. Note that unlike the previously mentioned techniques, no phase information is available, and often only the $|S_{11}|$, $|S_{12}|$ amplitude terms are available, not

$|S_{21}|$ or $|S_{22}|$. Poelman (1975) investigated the use of video addition to improve the detection of aircraft targets in noise; see Figure 5. Video addition was found to be advantageous because the target model used exhibited decorrelated Swerling I or II scintillation in each video channel. The target model also assumes some degree of cross–polar echo, $0 < F \leq 1$, where $F = \sigma_{cross}/\sigma_{like}$. The results of Poelman's theoretical and simulated data investigation are as follows. The technique of video addition can provide several decibels of improvement in the detectability factor over conventional single–channel detection when the number of hits is small ($n \leq 3$), $P_d \geq 0.8$, and the factor $F \geq 0.2$ in a Gaussian noise background. For targets with zero cross–polar component ($F = 0$) the maximum loss in detectability using this technique is 1 dB ($P_{fa} = 10^{-6}$, $P_d = 0.5$ and 0.8, $N \leq 20$).

A similar study has been completed using both measured and simulated data (Stewart, 1980). In this study, the like– and cross–polar echoes from various aircraft targets and ground clutter were measured from a ground–based X–band radar. It was found that the aircraft targets had roughly equal like– and cross–polarized echoes for circular polarization but quite weaker (about 9 dB) cross–polar echoes for linear polarizations. Additionally, the circularly polarized returns were almost uncorrelated (as described earlier by Poelman (1975)). For ground clutter, on the other hand, the cross–polarized echoes were generally about 6 dB weaker than like–polarized for linear polarizations, and about 2 dB stronger using circular polarizations. Using this data, Stewart concluded that dual circularly–polarized receivers improve the detection of aircraft in ground clutter, while at the same time dual linear polarization does not afford an improvement in target–to–clutter ratio, since the like polarized target echo is quite weak.

11.3.5.3   ICE FEATURE DETECTION

A different application involves the detection of various ice features in arctic regions. This detection has proved difficult using conventional non–coherent marine radars. The ice features of interest are those that pose a danger to shipping and to navigation, namely icebergs and their fragments (bergy bits and growlers) in open water, and multi–year sea ice floes in first year ice. Lewis, et al. (1985) have studied the non–coherent X–band echo behavior of icebergs and their fragments in open water. They have found that

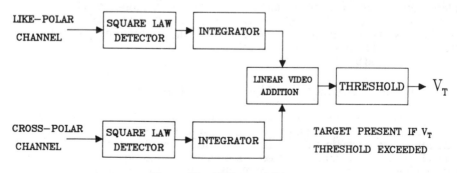

Figure 5   Video addition.

both ice targets and sea clutter tend to depolarize the incident horizontal polarization by the same degree, but that received like– and cross–polarized echoes have independent scintillation characteristics. Thus, improved detection of ice targets could be achieved by processing the dual–polarized receive echoes jointly.

The detection of icebergs or multiyear sea ice in full ice cover using ground based non–coherent radar has been investigated by Haykin, et al. (1985). In this case, the background clutter is viewed as the backscatter from first–year ice cover. In particular, experimental results using receivers at cross polarizations at Ka–band (35 GHz) demonstrated an improvement in target–to–clutter ratios of 10 to 12 dB for icebergs and 7 to 9 dB for multiyear ice. Similarly, at X–band an improvement of 6 to 7 dB for icebergs was measured. Though limited data was obtained, Ka–band also demonstrated an improved target–to–clutter ratio when a cross–polarized receiver was employed.

The above two studies (Lewis, et al. (1985) and Haykin, et al. (1985)) have demonstrated that improved detection of hazardous ice objects such as icebergs and their fragments as well as multiyear sea ice can be achieved by employing dual polarized receiver channels on a conventional non–coherent marine radar.

An operational trial of a ship–borne dual–polarized non–coherent X–band radar is described by Lewis and Currie (1987a). The ship trials were conducted using an Arctic ice breaker travelling from Belgium to Canada during the months of November and December, 1986. During this voyage various ice targets were encountered and documented, including complete first–year ice cover, with and without icebergs and their fragments, as well as multiyear sea ice. The average measured normalized radar cross sections, σ_o, of the various ice targets encountered during these trials compared favourably with the normalized radar cross sections measured earlier (Lewis, et al. (1985) and Haykin, et al. (1985)) using shore–based radar. Moreover, when the gain of the cross–polarized channel was properly adjusted, the PPI displayed only the desired ice targets with no false targets and no known target misses.

A thorough review of the many aspects of ice detection radar theory and performance evaluation can be found in Lewis, et al (1987b). The physical properties of ice are detailed as they relate to their radar scattering behavior, followed by the theory of surface–based ice surveillance radar. The remainder of the book describes extensive measurement and analysis of multifrequency polarization radar returns from various ice features over a period of 6 years. Interestingly, the final proposed solution to both the iceberg/fragment and multiyear–ice detection problems rely on using cross–polarized echoes obtained from a dual–polarized radar.

11.3.5.4 POLARIMETRIC RADAR NAVIGATION

Radar navigation along a confined waterway (e.g., the St. Lawrence Seaway, the Mississipi River) using passive shore–based retro–reflectors is yet another important, successful application of radar polarimetry. Basically, the clutter echo on a linear cross–polarized channel is weaker (typically by 5 to 20 dB) than the corresponding linear like–polarized echo. Hence, by exploiting this reduction in clutter induced cross–polarized echo, the retro–reflector–to–clutter ratio may be improved by 5 to 20 dB. The polarimetric radar navigation system thus employs a dual–polarized radar antenna on receive.

The wide–angle, passive, polarization–modifying (polarimetric) reflectors make the above radar navigation system possible. These reflectors must change incident linear horizontal polarization to linear vertical or vice versa, and must do so over a wide range of incidence angles in azimuth and elevation. A reflector possessing all the above attributes is described in the patent application, Macikunas, et al. (1984) and in Haykin, et al. (1985).

An experimental test of the PRAN system employing the above–mentioned polarimetric reflectors was conducted from 1984 to 1986 and is described in Haykin, et al (1986). The above experimental setup used a conventional X–band non–coherent marine radar fitted with a dual–polarized antenna, feed and extra receiver in an operational environment consisting of sea clutter, natural and man–made clutter and shipping traffic. It was found that radar echoes from the special polarimetric reflectors could be detected very easily in either a benign natural clutter setting, or amongst the hostile man–made clutter. Not surprisingly, conventional like–polarized receiver echoes of standard trihedral retroreflectors were not discernible at all in the man–made clutter environment. These experimental results prove the validity of the PRAN concept and the utility of the special polarimetric reflectors.

The application of radar polarimetry to the detection of radar buoy targets (conventional – not polarimetric as described above) in sea clutter is reported by Long (1980). Long utilizes the fact that sea clutter echoes from reflector targets usually have roughly equal echo power on both linear transmit/receive polarizations (HH or VV) while sea clutter echo does not. By transmitting a signal with equal horizontal (H) and vertical (V) polarization components, both H and V polarized echo components can be received simultaneously. The target/ clutter discrimination is realized by comparing the ratio of received H and V echo amplitudes to unity. For targets, the ratio will be near unity, and for clutter or interference the ratio will be either very small or very large. An improvement of several dB in target to clutter ratio is expected using this technique.

11.3.5.5 THE PSEUDO–COHERENT DISCRIMINANT

Another method of improving the detection and discrimination capabilities of a radar is to use the relative phase between the two orthogonally–polarized received waves, this is known as the pseudo–coherent discriminant (PCD). The technique is pseudo–coherent in that a coherent transmitter or a transmitted phase reference are not required, hence, a conventional non–coherent radar can be utilized. The measurement of the pseudo–coherent discriminant is illustrated in Figure 6.

First, both channels are hard–limited to remove amplitude information; next the two channels are fed to a phase detector — a mixer with one channel acting as the local oscillator. The output, V_ϕ, is the sine of the relative phase between the two channels. The sine of the polarimetric phase

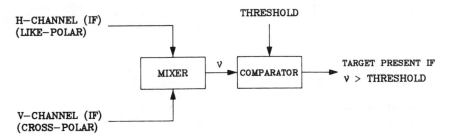

Figure 6   Pseudo–coherent discrimination.

is then applied to a comparator with a threshold level for target present or absent decision.

Another form of the above detector is called the pseudo–coherent discriminant/dot product.   This form of detector does not discard the amplitude component of the two signals; rather, the detection can be looked upon as the dot product between the H–channel, $H\angle\phi_H$ and V–channel $V\angle\phi_V$;   see Figure 7, note the output voltage $v(t) = H(t) \cdot V(t) \; \sin(\phi_H(t) - \phi_V(t))$.   Stovall (1978) undertook a study of the pseudo–coherent discriminant for fixed retroreflector targets in Gaussian noise and compared the pseudo–coherent detector to Marcum square–law detection.   Stovall's results indicate that the pseudo–coherent discriminant could approach the performance of the Marcum square–law detector provided that the number of integrated pulses, N, was large (> 20).   Obtaining the large number of independent samples in a noise background is easy.   For clutter, independent samples can only be obtained through frequency–agility, and the number of such samples is limited by the time–bandwidth product.   Moreover, such large time–bandwidth systems must maintain their phase coherence (including antenna, feed, r.f. hardware) over the entire agile bandwidth.   Stovall also notes that typical clutter will produce a bias in the ϕ_{rel} term because clutter will likely not completely depolarize the transmitted waveform.   This effect should degrade the PCD performance – further study is recommended.   Lastly, Stovall notes that despite the PCD's optimum performance without an adaptive threshold, the Marcum square–law detector operates as well with only one channel as the PCD with two channels (intuitively, we expect some improvement if two channels are used).

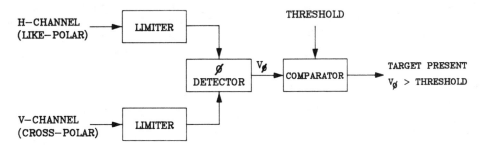

Figure 7   Pseudo–coherent discriminant/dot product.

Vachula and Barnes (1983) have compared several detection statistics (mentioned earlier), including Stokes parameters, marcum square–law detection, and the PCD. For the cases studied (N = 10, and 30, P_{fa} = 0.067 and 0.000693) the PCD was found to have 2.5 to 3.0 dB worse performance than the best Stokes–parameter–based statistics, and roughly equal performance to the Marcum square–law detector.   This result is somewhat misleading since the square–law and Stokes–parameter–based statistics are calculated for a Swerling model fluctuating target, and the PCD result is for a stationary target.   A fluctuating target could degrade the performance of the PCD by a factor of 1 or 2 dB more, making it even less appealing.

The theoretical models and simulated performance for both the PCD and the PCD/dot product have been derived by Echard, et al. (1981).   This study considers a non–fluctuating target in Rayleigh distributed clutter (both channels independent).   The PCD/dot product detector was found to have a 3 dB advantage over square–law detection for single pulses, and as little as a 1.6 dB advantage when very many pulses are integrated (> 100).   Again, the PCD performed worse than ordinary square–law detection but required no CFAR threshold setting which is required by both the Marcum square–law detection and PCD/dot product detector.

11.3.6   DIFFERENTIAL REFLECTIVITY

The shapes of hydrometeors falling in the atmosphere are usually nonspherical, which results in polarization–dependent scattering properties. Moreover, there is evidence to suggest that hydrometeors fall with a preferred orientation.   Accordingly, we may use polarimetric radar techniques that rely on the differential scattering properties between horizontally and vertically polarized waves (Seliga and Bringi, 1976).   In this paper, the differential scattering properties of classes of hydrometeors at linear orthogonal polarizations are exploited for radar measurements of precipitation.   Specifically, Seliga and Bringi advanced the use of differential reflectivity, defined by

$$Z_{DR} = 10 \log_{10} \frac{Z_H}{Z_V}$$

where Z_H and Z_V are the horizontal and vertical radar reflectivity factors, respectively.   The differential reflectivity factor Z_{DR} is independent of radar constants for equal system response at both polarizations.   Indeed, it depends only on D_o, the median diameter of the assumed raindrop size distribution. Hence, Z_{DR} should be determined precisely through the use of relative power measurements, thereby yielding D_o directly.   The differential reflectivity Z_{DR} may be combined with differential phase shift to improve the accuracy of rain fall rate estimation (Seliga and Bringi, 1978).   Furthermore, polarization diversity may be used to develop remote methods of hydrometeor identification in storms as evidenced by recent results of successful hail detection. Sachidananda and Zrnic (1986) describe a signal processing scheme for a radar with alternating polarization, which may be used to measure the differential reflectivity and differential propagation phase shift.

11.3.7 CORRECTIONS TO THE SCATTERED WAVE

It is often desirable to make two types of corrections to received polarimetric waves, one involving the scattering process itself, the other dealing with the transmission/reception system. The first form of correction deals with the application of inverse theories in optics. Boerner (1981) discusses how such techniques could be applied in radar scatter depending on the relative object size to wavelength. From a processing standpoint, a more important consideration is the correction of the polarization–errors in the transmission/ reception system and possibly in the propagation medium. These errors are caused by the cross–coupling of energy from one channel to the other orthogonal channel and vice versa. Imperfect transmit/receive antennas, and their feeds can lead to significant cross–coupling. Uncorrected, cross–coupling leads to significant degradation in the performance of most polarimetric processing schemes outlined earlier.

Earlier, in Section 11.2.1, the measured scattering matrix was given in Equation (4) as the product of antenna, propagation and target scattering effects. It is desirable to cancel the effects of both antenna and propagation before the polarimetric processing. Since propagation effects are most pronounced for long–range and high elevation earth–space paths, they will be assumed to be negligible for short terrestrial radar paths. The remaining errors in the antenna are a combination of cross–polar characteristics of the antenna, reflector, associated feeds, and orthomode transducers (if used). Several techniques to correct for the effect of antenna/feed are described below.

McCormick and Hendry (1979) describe various phenomenon associated with the measurement of precipitation depolarization data and the correction of polarization errors. The system considered utilizes a monostatic dual– orthogonal circularly–polarized antenna with ability to transmit either polarization while receiving both orthogonal polarizations. A network of phase shifters and attenuators is provided to adjust antennas for greatest polarization purity using calibration signals. Theoretical derivations are given relating the received amplitudes, respective channel gains and offset geometry to polarization errors. A later paper by McCormick (1981) elaborates on the above relations giving inverse transforms based on the measured polarization errors to restore polarization orthogonality. McCormick notes that these corrections can only be valid over a given part of the antenna angular response. This last problem arises since all antennas exhibit markedly different cross–polar characteristics in amplitude and phase over their beams (particularly in the sidelobes).

The problem of cross–coupling dual–orthogonal receive channels also occurs in the field of microwave radio data communications. Amitay and Salz (1984) apply linear least–mean squares equalization to two dual–polarized radio channels. In particular, the effects of polarization errors due to the antennas and their feeds could be practically eliminated. In their paper a channel model is derived based on a 2×2 square matrix composed of co–polarized terms on the diagonal, and cross–polar (leakage) terms in the off–diagonal, all as a function of frequency. Based on this model and other parameters, optimum receiver and transmitter structures are given as well as the expected mean square errors. The least–mean–square error equalization was found to be very successful at suppressing significant antenna and feed cross–polar coupling.

Tseng and Lee (1982) derive a technique to restore polarization orthogonality in satellite channels. Their technique unfortunately requires extra cross– polarized pilot transmitters and receivers with their own antennas, making it useless for conventional radar systems.

11.4 CONCLUSIONS

In this chapter we have presented a technical review of polarization as a radar discriminant. The polarization nature of a radar wave and the mathematical representation of polarized waves were described. In particular, target and clutter discriminants based on optimal polarization, Stokes vector/matrix, polarization vector processing and Hermitian coherency matrix were compared for coherent or partially coherent radars.

The use of a polarimetric radar requires increased equipment complexity, compared to a conventional radar. In return, the use of polarization provides the potential for improved detection or classification of radar targets of interest. Indeed, there are situations where the use of polarization as a radar discriminant offers the only means of improved or satisfactory performance. Some potential application areas include: aircraft detection, radar remote sensing, marine/coastal radar, surface–based radar, and polarimetric radar navigation along confined waterways.

The mathematical representation of polarized waves may be subdivided into two areas: the representation of partially–polarized (time–varying) waves and the representation of completely–polarized (non time–varying) waves. Completely–polarized waves could be conveniently represented using a polarization vector. Any two polarization vectors may be related by a transformation known as the polarization scattering matrix (PSM). The PSM could also be transformed by other matrices representing the polarization characteristics of the receiving antenna, propagation medium, etc. An eigenvalue analysis of the above PSM is found to yield the optimal (characteristic) polarizations of the target. The 2 or 3–dimensional target polarization data cannot be easily displayed without the use of some form of projection or chart. To this end, the Poincare sphere, modified polarization chart, the 'Okishi' chart, and the θ–τ–intensity chart (Zebker, et al, 1987) may be utilized.

In the more general case of partially–polarized waves the description and graphic representation of the wave becomes more complicated. The Stokes vector is defined as a means of describing the partially–polarized wave and the Mueller matrix is given as a transformation between two partially–polarized waves. Mathematical relationships between the Mueller matrix and the PSM were given – this is useful since both were used to describe a polarization transformation. In addition to the Mueller matrix, two other partially–polarized wave representations may be considered, namely: the coherency matrix and the Hermitian coherency matrix (Cloude, 1986a, 1986b).

Many of the previously described mathematical representations of polarized waves can be used directly as polarization discriminants. For example, an enhanced type of radar display can be created using the echo amplitude for intensity and the polarimetric (relative) phase for colour. This enhanced display has been shown to provide effective discrimination between different target classes in remote sensing radar images. The optimal polarizations could be used to classify or detect different types of polarization–modifying targets in clutter or noise; for example, ships could be distinguished from polarization–modifying targets such as dihedrals or trihedral twist grid reflectors. This could be achieved through the use of null polarizations themselves or through the use of the variance of the optimal polarizations or the logarithm of the transformation parameter, ρ. The matrix difference measure of the polarization covariance matrix is also capable of better detection performance than Marcum square law detection of horizontal amplitude echo under most conditions.

Another powerful approach described is polarization–domain vector processing. Using this technique, the detection performance of aircraft surveillance radars could be improved by 9 dB in rain clutter, or the ground clutter could be attenuated by 14 dB without sacrificing any additional moving target indicator (MTI) performance, or a jammer could be attenuated by about 14 dB at a Target/Jammer ratio of approximately 0 dB. Two structures, the multinotch logic–product filter and the adaptive polarization canceller, were described to achieve this filtering operation.

The use of the Stokes vector approach to describe polarized waves leads to yet another class of detection statistics. These statistics include the likelihood ratio (Stokes parameters $g_0 + g_3$) and the Mueller matrix elements.

Dual–polarized radars which are non–coherent reduce the amount of information available to the receiver. Under such conditions, it is often possible to perform some type of sub–optimum polarization processing on the radar echo. For example, the use of video addition of the two orthogonally–polarized echo signals yields greatly improved detection of aircraft in noise since the aircraft echo scintillates almost independently in both channels. Also, the use of a linear cross–polarized receiver may improve both the detection of various ice targets (increase T/C ratio by up to 12 dB) and the discrimination of multi year sea ice from a first–year sea ice background by up to 9 dB. Described also was a system that could improve the visibility of shore–based polarimetric retroreflectors used in marine navigation through the use of a dual–polarization radar. Another non–coherent radar alternative considered was the pseudo–coherent discriminant (PCD) and the PCD/dot product. These discriminants, particularly the PCD/dot product, were found to be up to several dB better than the Marcum square law detector when the number of hits is low. Lastly, there is a need to correct the received dual–polarization echo signals since imperfections in the radar antenna/feed and propagation effects can cause a cross–coupling of signal from one polarization to the other degrading the detection/discrimination performance.

11.5   REFERENCES

Amitay, N. and J. Salz (1984). "Linear Equalization Theory in Digital Data Transmission Over Dually Polarized Fading Radio Channels", AT&T Bell Systems Technical Journal, V. 63, No. 10, pp. 2215–2259.

Bickel, S.H. and R.H.T. Bates (1965). "Effects of Magnetoionic Propagation on the Polarization Scattering Matrix", Proc. IEEE, p. 1089.

Boerner, W.M., M.B. El–Arini, C.Y. Chan and P.M. Mastoris (1981). "Polarization Dependence in Electromagnetic Inverse Problems", IEEE Trans. Antennas and Prop., V. AP–29, No. 2, pp. 262–271.

Born, M. and E. Wolf (1980). Principles of Optics, 6th Ed., Pergammon Press, Oxford, England.

Carver, K.R., C. Elachi, and F.T. Ulaby, (1985). "Microwave Remote Sensing from Space", Proc. IEEE, V. 73, pp. 970–996.

Cloude, S.R. (1983). "Polarimetric Techniques in Radar Signal Processing", Microwave Journal, V. 26, No. 7, July, pp. 119–127.

Cloude, S.R. (1985). "Target Decomposition Theorems in Radar Scattering", Electronics Letters, V. 21, No. 7, pp. 22–24.

Cloude, S.R. (1986a) "Group Theory and Polarization Algebra", Optik, V. 75, pp. 26–36.

Cloude, S.R. (1986b). "Polarimetry: The Characterization of Polarization Effects in EM Scattering", Ph.D. Thesis, Univ. of Birmingham, England.

Copeland, J.R. (1960). "Radar Target Classification by Polarization Properties", Proc. IRE, pp. 1290–1296.

Corsini, G., E. Dalle Mese, M. Mancianti, and L. Verrazzani (1984). "Classification of Radar Targets: An Overview", Proc. 1984 Int. Symp. on Noise and Clutter Rejection in Radars and Imaging Sensors, Tokyo, Japan, pp. 65–70.

Daley, J.C. (1980). "Radar Target Discrimination Based on Polarization Effects", (NRL) Workshop on Polarimetric Radar Technology, Huntsville, AL, pp. 185–213, (published 1981).

Daley, J.C. (1982). "Radar Target Classification Based on Polarization Matrix", 2nd Annual Workshop on Polarimetric Radar Technology, Huntsville, AL (published 1983).

Déschamps, G.A. (1951). "Part II – Geometric Representation of the Polarization of a Plane Electromagnetic Wave", Proc. IRE, pp. 540–544.

Echard, J.D., E.E. Martin, D.L. Odom and H.B. Cox (1981). "Discrimination Between Targets and Clutter by Radar", Georgia Tech., Report CIT/EES–A–2230 FTR.

Emmons, G.A. and P.M. Alexander (1983). "Polarization Scattering Matrices for Polarimetric Radar", U.S. Army Missile Command, Redstone Arsenal, Alabama, Rep. AD–A 147–373.

Ezquerra, N.F. (1984). "Radar Target Recognition: A Survey", Proc. Inter. Conf. Radar 84, Paris, France, pp. 281–285.

Fossi, M., M. Gherardelli, D. Giuli, L. Piccini, and G. Ponziani (1984). "Experimental Results on Discrimination of Radar Signals by Polarization", Int. Symp. on Noise and Clutter Rejection in Radars and Imaging Sensors, Tokyo, Japan, pp. 501–505.

Gatley, A.C. (Jr.), D.J.R. Stock and B. Ru–Shao Cheo (1968). "A Network Description for Antenna Problems", Proc. IEEE, V. 56, No. 7, pp. 1181–1193.

Giuli, D., M. Gherardelli, and E. Dalle Mese (1982). "Performance Evaluation of Some Adaptive Polarization Techniques", Inter. Conf. Radar, London, U.K., pp. 76–81.

Giuli, D., M. Fossi, M. Gherardelli (1985). "A Technique for Adaptive Polarization Filtering in Radars", Conf. Proc. IEEE Inter. Radar Conference, Arlington, VA, U.S.A. pp. 213–219.

Giuli, D. (1986). "Polarization Diversity in Radars", Proc. of IEEE, V. 74, No. 2, pp. 245–269.

Grasso, G. and S. Pardini (1975). "Polarization Effect on Multipath for Extended Targets", IEEE Trans. on Aerospace & Elect. Systems, V. AES–11, No. 3, pp. 316–320.

Graves, C.D. (1956). "Radar Polarization Power Scattering Matrix", Proc. IRE, pp. 248–252.

Haykin, S., A. Macikunas, and T. Greenlay, (1985). "Polarimetric Radar for Precise Navigation Along Confined Waterways", Proc. European Microwave Conf., Paris, France, pp. 137–142.

Haykin, S., B.W. Currie, E.O. Lewis, and K.A. Nickerson, (1985). "Surface–Based Radar Imaging of Sea Ice", Proc. IEEE, V. 73, pp. 233–251.

Haykin, S., R. Cho, and T. Greenlay, (1986). "Preliminary Experimental Results of Polarimetric Marine Radar for Accurate Navigation", Conf. Proc. IEEE Montech '86, Montreal, Canada, pp. 65–68.

Holm, W.A. (1984). "Polarization Scattering Matrix Approach to Stationary Target/Clutter Discrimination", Proc. Int. Conf. on Radar, Paris, France, pp. 461–465.

Huynen, J.R. (1965). "Measurement of the Target Scattering Matrix", Proc. IEEE, pp. 936–946.

Huynen, J.R. (1970). "Phenomenological Theory of Radar Targets", Ph.D. Thesis, Drukkeriz Bronder–Offset, NV Rotterdam.

Huynen, J.R. (1982). "A Revisitation of the Phenomenological Approach with Applications to Radar Target Decomposition", Comm. Lab., Dept. of Inf. Eng., Univ. of Ill., EMID–CL–82–05–18–01.

Kennaugh, E.M. (1981). "Polarization Dependence of RCS — A Geometrical Interpretation", IEEE Trans. Ant. and Prop., V. AP–29, No. 2, pp. 412–413.

Ko, H.C. (1962). "On the Reception of Quasi–Monostatic, Partially Polarized Radio Waves", Proc. IRE, pp. 1950–1957.

Kostinski, A.B., and W.M. Boerner, (1986). "On Foundations of Radar Polarimetry", IEEE Trans. Ant. and Prop., V. AP–34, pp. 1395–1404.

Kostinski, A.B., and W.M. Boerner (1987). "On the Polarimetric Contrast Optimization", IEEE Trans. Ant. and Prop., V. AP–35, pp. 988–991.

Lammertse, H. and E. Goldbohm, (1984). "Research on Polarization Properties of Seaclutter", Proc. of Int. Symp. on Noise and Clutter Rejection in Radars and Imaging Sensors, Tokyo, Japan, pp. 111–116.

Lewis, E.O., B.W. Currie, and S. Haykin (1985). "Effects of Polarization on the Marine Radar Detection of Icebergs", Proc. IEEE Int. Radar Conference, Versailles, France, pp. 253–258.

Lewis, E.O., and B.W. Currie (1987a). "Arctic Ship Trials of a Dual–Polarized Ice navigation Radar", Int. Conf. on Radar '87, London, U.K., pp. 404–408.

Lewis, E.O., B.W. Currie, and S. Haykin (1987b). Detection and Classification of Ice, Research Studies Press, Chichester, England.

Long, M.W. (1980). "New Type Land and Sea Clutter Suppressor", Proc. IEEE Int. Radar Conference, pp. 62–66.

Lowenschuss, O. (1965). "Scattering Matrix Application", Proc. of IEEE, pp. 988–992.

Macikunas, A., S. Haykin, and T. Greenlay (1984). "Trihedral Radar Reflector", Canadian Patent Applied for Nov. 22, 1984, U.S. Patent Applied for Nov. 1985, Oakville, Canada.

Macikunas, A., and S. Haykin, (1986). "Radar Polarimetry for Improved Detection of Targets in Clutter", Conf. Proc. IEEE Montech '86, Montreal, Canada, pp. 61–64.

McCormick, G.C. and A. Hendry (1979). "Techniques for the determination of the polarization properties of precipitation", Radio Science, V. 14, No. 6, pp. 1027–1040.

McCormick, G.C. (1981). "Polarization Errors in a Two–channel System", Radio Science, V. 16, No. 1, pp. 67–75.

Metcalf, J.I. (1984). "Interpretation of the Auto–covariances and Cross–covariance from Polarization Diversity Radar", Air Force Geophysics Lab., Mass., U.S.A., report AD–A144–907.

Mieras, H. (1983). "Optimal Polarizations of Simple Compound Targets", IEEE Trans. Ant. and Prop., V. AP–31, No. 6, pp. 996–999.

Murza, L.P. (1978). "The Noncoherent Polarimetry of Noiselike Radiation", Radio Engineering and Electronic Physics, V. 23, pp. 57–63.

Nathanson, F.E. (1975). "Adaptive Circular Polarization", Proc. IEEE Int. Conf. on Radar, Washington, DC, pp. 221–225.

Nespor, J.D., A.P. Argawal and W.M. Boerner (1984). "Development of a Model–Free Clutter Descriptor Based on a Coherency Matrix Formulation", IEEE Ant. and Prop. Int. Symp. Digest, pp. 37–40.

Okishi, T. (1986). "A Planar Chart Equivalent to Poincaré Sphere for Expressing State–of–Polarization of Light", Journal of Lightwave Technology, V. LT–4, pp. 1367–1372.

Poelman, A.J. (1975). "On Using Orthogonally Polarized Noncoherent Receiving Channels to Detect Target Echoes in Gaussian Noise", IEEE Trans. on Aerospace and Electronic Systems, V. AES–11, pp. 660–663.

Poelman, A.J. (1976). "Cross–correlation of Orthogonally Polarized Backscatter Components", IEEE Trans. on Aerospace & Electronics Systems, V. AES–12, no. 6, pp. 674–682.

Poelman, A.J. (1980). "A Study of Controllable Polarization Applied to Radar", Conf. Proc. of Military Microwaves, London, U.K., pp. 389–404.

Poelman, A.J. (1981). "Virtual Polarization Adaptation," IEE Proceedings, V. 128, No. 5, Pt. F, pp. 267–270.

Poincaré, H. (1892). Theorie Mathematique de la Lumiere, Georges Carre, Paris, France.

Pickard, J.T. (1985). "On the Detection Theory of Autocoherence", IEEE Trans. Info. Theory, V. IT–31, No. 1, pp. 80–90.

Raven, B. (1985). "On the Null Representation of Polarization Parameters", Inverse Methods in Electromagnetic Imaging, W. Boerner, et al. Editors, Reidel Publ., Dordrecht, Holland, pp. 629–641.

Russell, R.F., F.W. Sedenquist, and C.L. Broten (1987). "Radar Target Characterization Utilizing a Polarimetric Technology Sensor", Proc. Int. Conf. on Radar, London, U.K. pp. 365–369.

Sachidananda, M., and D.S. Zrnic (1986). "Characterization of Echoes from Alternately Polarized Transmission", Cooperative Institute for Mesoscale Meteorological Studies, Report No. 71, Norman OK, USA.

Seliga, T.A., and V.N. Bringi (1976). "Potential use of radar differential reflectivity measurements at orthogonal polarizations for measuring precipitation", J. App. Meteorology, V. 15, pp. 69–76.

Seliga, T.A, and V.N. Bringi (1978). "Differential reflectivity and differential phase shift: Applications in radar meteorology", Radio Science, V. 13, pp. 271–275.

Sinclair, G. (1950). "The Transmission and Reception of Elliptically Polarized Waves", Proc. IRE, pp. 148–151.

Stewart, N.A. (1980). "A Study of Polarization Characteristics of Targets and Clutter", Conf. Proc. of Military Microwaves, London, U.K. pp. 405–410.

Stock, D.J.R. and H. Trogus (1984). "Polarization Utilization in Large Radars", Proc. Int. Conf. on Radar, Paris, France, pp. 466–471.

Stokes, G. (1852). "On the Composition and Resolution of Streams of Polarized Light from Different Sources", Trans. Cambridge Phil. Soc., V. 9, Pt. 3, pp. 399–416.

Stovall, R.E. (1978). "A Gaussian Noise Analysis of the Pseudo–coherent Discriminant", M.I.T. Lincoln Laboratory, Report ESD–TR–78–395, Lexington, Mass.

Tseng, F.T. and L.S. Lee (1982). "The PQ–Plane Approach for Restoring the Polarization Orthogonality in a Multiple Access Satellite System", Globecom Conference Rec., Miami, FLA, pp. 1284–1288.

Vachula, G.M. and R.M. Barnes (1983). "Polarization Detection of a Fluctuating Radar Target", IEEE Trans. Aerospace and Electronic Systems, V. AES–19, No. 2, pp. 250–257.

Van Zyl, J.J. (1987). "On the Optimum Polarizations of Incoherently Reflected Waves", IEEE Trans. Ant. and Prop. V. AP–35, pp. 818–825.

Wiener, N. (1927–1929). "Coherency Matrices and Quantum Theory", J. Math and Phys., V. 7, pp. 109–125.

Wiener, N. (1930). "Generalized Harmonic Analysis," Acta Math, V. 55, pp. 117–276.

Wiesbeck, W. and S. Riegge (1987). "Measurement of the Complex RCS–matrix for Small Metallic and Dielectric Objects", Proc. Int. Conf. on Radar, London, U.K., pp. 360–364.

Zebker, H.A., J.J. VanZyl, and D.N. Held (1987). "Imaging Radar Polarimetry from Wave Synthesis", J. of Geophysical Res., V. 92, pp. 683–701.

INDEX